Modern Astrophysics Meets Engineering

On the Reformation of Physics

Modern Astrophysics Meets Engineering

On the Reformation of Physics

Mathias Hüfner

The most dangerous of all worldviews is the worldview of the people, who did not look at the world. - Alexander von Humboldt -

In the GDR we demanded a worldview without being allowed to look at the world. Today we are allowed to look at the world, but many people remain blind with seeing eyes.

Bibliographische Information der Deutschen Nationalbibliothek:
Die Deutsche Nationalbibliothek verzeichnet diese Publikation
in der Deutschen Nationalbibliographie; detaillierte bibliographische
Daten sind im Internet über htpps://www.dnb.de abrufbar.

Herstellung und Verlag
BoD – Books on Demand, Norderstedt

ISBN 978-3-75192-018-6

Table of Content

Preface of the English Edition

My guiding idea: If you have no clear idea of a natural phenomenon, then no mathematical description helps to gain more clarity. But you can hide better your ignorance.

The book is an extended translation of the German book "*Astrophysik trifft auf Ingenieurwissenschaften*". As the engineering sciences celebrate success after success in space technology, academic astrophysics is sinking more and more into the arms of the faith in medieval scholasticism. There are two fundamentally different ways of thinking. While astrophysics is trapped in belief, the engineer's mindset is an analytic, feasibility mindset.

Let us take a few examples:

1. No engineer would come up with the idea of car crash testing to unravel the constructive secrets of competition. Physicists are doing crash tests with atoms to explore the interior of the atomic nucleus.
2. No engineer would want to determine the initial state of a design from the blueprint. He uses the knowledge of manufacturing technologies. Physicists draw from the image of today's cosmos to the initial state without ever having understood the developmental stages of the cosmos.
3. No engineer would use more resources than necessary. Physicists design 9-dimensional spaces to move maximum 2-dimensional objects.
4. No engineer would build a machine that can handle both large and microscopic objects. Physicists want to develop a unified theory for the macrocosm and microcosm.
5. No engineer would trust his mind without having made an experiment and having subjected the results to a discussion of al-

ternatives. Physicists make experiments to validate their beliefs, not to falsify.

The analytical thinking of the engineers we owe the successes of satellite technology in the exploration of the universe. We should also trust the analytic abilities of the engineers more than the fantasies of the astrophysicists when interpreting the measurement results.
The physical fantasies are massively promoted by the power interests of the faith representatives of the Catholic Church, without the public being aware of it. This claim to power is formulated in the encyclical Pascendi Dominici gregis of Pope Pius X. in 1907 and is still valid today. It is enforced worldwide by Jesuit schools and follows the mainstream of physicists. According to the specifications of this encyclical against modernity, no other science has become so successful a 'maidservant of theology' as astrophysics. For this purpose, the Standard Model of Cosmology and the Standard Model of Particle Physics were designed around a divine creation called the Big Bang. Einstein's theory of relativity is the basic building block of this model, which was appropriated by G. Lemaitre and developed by subsequent generations of astrophysicists. Pope Pius XII even made the Big Bang model the foundation of the Catholic faith, making it hard for believing scholars to part with.

Mathias Hüfner, Jena on 04/22/20

1 Why this Book?

Homo sapiens is a thinking man. Anyone who thinks can be wrong because an inductive conclusion must be limited and the limit is initially unknown, but only a fool insists on his error.

These days, many people write books or have written books by Ghost-writers. Most believe they should disclose their lives to the world as if everyone isn't sufficiently occupied with their own lives. Because I don't want to lose you, dear reader, here are just a few sentences about myself to convince you that I know what I am writing about and that I am seriously seeking knowledge, because faith is given to us, but for knowledge we must be pursued and knowledge begins where the faith stops. The path to knowledge goes beyond looking at the world and questioning what has been learned. Knowledge is not static. New findings from space exploration must be integrated into our knowledge building. Renovation measures are not enough. This is very painful for those who have settled comfortably in the affected rooms of this knowledge building. Their inhabitants will defend themselves with all their power and hostility and slander are not excluded.

In the seventies of the last century I studied physics in Leipzig. I struggled with two conceptual problems. The first concerned the theory of relativity and the second was the duality of light as a wave and a particle. After completing my professional career as an engineer, I turned back to these two questions, remembering the question marks I left in my textbooks as marginal notes, questions that I did not dare to ask as a student, and finally let me in different direction career development. Originally my main concern was to fill the days of my retirement with a meaningful activity and at the time I had no idea where the answers to these two questions would lead me. In the last decade, I've learned so much about physics that I've developed a whole new perspective on this subject. I should explain something. As a physicist at a university, you live under a system of protection. If you show good behavior, you will be accepted into the rope team and your career will be assured. Moving out of the mainstream is a risk that should not be underestimated. As an engineer, you have to prove yourself in the market. The easy path is blocked by patents. There's nothing to gain with a team of riders. Engineers are free climbers. In other words, for an engineer, there can be no alternative solution. He must always think of new ways to get to the destination. Of course, there are now also in industry attempts of protectionism. One example is the German car industry with its diesel scandal. Ultimately, however, such attempts fail in the market. My conclusion was this: never follow the prescribed reasoning, but look for another way to come to a solution. If you then arrives the same result, the given solution is acceptable. But if you come to a different result, ask for the cause. In the following book, we will look at the most diverse aspects of why engineering in physics comes to different conclusions than the so-called 'Modern Physics', which believes to be based on pure reason, on mathematics. We will gradually destroy this illusion.

Physics not only seems to have crossed the boundaries of our discernible world, but also overwhelmed our minds. Here we want to explain why fantasy and reality are so mixed up in today's physics, what role mathematics plays in this and how it can be unraveled. In the meantime, the amount of substance on my website has grown so much that I have to give an interested layman a guideline as I write something like that not only for my own understanding. However, finding such a guideline is only possible if you yourself are standing far enough above the material to be able to work out the essential structures of the material. So it's not about a textbook in the traditional sense, but rather about sustainable physical concepts that are consistent with human experience and that can not be ignored if you go into unknown worlds.

Physics is the basis of science and, as such, it is based on observation and experimentation, but the interpretation of the results depends in an astonishing way on our beliefs. But that means that no matter what we measure, if we have a different belief, we get another physics.

Which then is true physics, Classical or Modern physics?

Their own faith would be defended like their own country, claims Hilton Ratcliffe. [1.01] He lives in South Africa and has not experienced the peaceful revolution in East Germany. If, for example, faith no longer corresponds to real life circumstances, it feeds doubt. Now, most people's living conditions have virtually nothing to do with Modern physics, and the public's interest in them is correspondingly weak. At least the TV series "Star Trek" may still remember some. There you could penetrate into the distant worlds of the cosmos, encountering space travelers with black holes, wormholes, and other fantastic adventures in distant galaxies, all of which has little to do with physics. The seductive message in this series is that we could leave our earth and seek a new earth if we destroyed the existing ones. Anyone looking around the world realizes that we are on the right track to global destruction.

It should be clear to everyone that leaving our planet remains an illusion for us, even though the arrival of planets in our solar system is now a reality for space probes. Unless human civilization is able to protect the Earth's environment, it will not be able to implement a viable biosphere in a spaceship. To grasp that, a dose of physics is quite useful. There are some people today, such as engineers and technicians, who are starting to push back against Modern physics because they find it contradictory without being able to pinpoint the causes of the contradictions. A scientific teaching, unlike a religion, must be consistent. When contradictions to reality occur, it's a sign that the descriptive theory is wrong and you have to rethink the whole thing from the start.

Physics is primarily a measuring science. Consequently, what you can not measure cannot be of interest to physics. The structure of physical knowledge is such that it is organized into mechanics, electrodynamics, optics, thermodynamics, atomic physics, nuclear physics, and so on. This division of knowledge works like the demarcation of principalities. In each one there is a professor who rules his field of knowledge as a sovereign prince. There are fixed demarcations and the princes do not mind whether the findings in one area are compatible with those of another. On the contrary, they strictly guard that against a stranger crossing. That does not always apply.

It occurs to me that there was a battle between a Prince of Quantum Theory, Leonard Susskind, and the Dark Lord of the Theory of Relativity, Stephen Hawking, when discussing whether the information was preserved in a black hole or not. This battle has gone down in history as the Black Hole War. Since this question could not be decided for a long time - as well as, after all, this can only be decided through an experiment - the opponents have meanwhile agreed that the information must hang on the event horizon, as if it had been tacked there. After all, Hawking has settled the dispute in early 2014 and declared that there is no event horizon and therefore no black holes exist.

It occurs to me that there was a battle between a Prince of Quantum Theory, Leonard Susskind, and the Dark Lord of the Theory of Relativity, Stephen Hawking, when discussing whether the information was preserved in a black hole or not. This battle has gone down in history as the Black Hole War. Since this question could not be decided for a long time - as well as, after all, this can only be decided through an experiment - the opponents have meanwhile agreed that the information must hang on the event horizon, as if it had been tacked there. After all, Hawking has settled the dispute in early 2014 and declared that there is no event horizon and therefore no black holes exist.

As ridiculous as the subject of the dispute is, it shows that there was at least one attempt to remedy a malady in physics. However, there are too many now that one can speak of a crisis in Modern physics. Even the well-known theoretician Lee Smolin devoted a whole book to this topic under the title " Trouble with Physics ", only it was not consistent. Physics is permeated with metaphysics[1]) and hardly anyone can say with certainty what is still physics and what is already metaphysics. So my concern was to find those boundaries. How can this happen ?

---

1 The term goes back to Aristotle and means what follows behind physics, the natural philosophy. These are the things that can no longer be sensed.

The key lies in mathematics and in the possibilities of its generalization. Mathematics is the most important tool for physics in order to come to accurate quantitative statements within the limits of measurement. In the last century, a mathematization of physics has taken place as never before. However, this was not as beneficial as you would have wished. Mathematicians have taken over physics, developed dubious theories such as relativity theory and string theory, and degraded experimental physics to prove these theories. »*If any experiment contradicts a beautiful idea, let us forget the experiments*«. This sentence pronounced by Dirac or Einstein - the sources contradict each other - illustrates the way of thinking that has been established for several generations in the circles of theoreticians of this physics, namely the overvaluation of the theory over experiment.

Since no one finds it necessary to clean up the physics in the main office, I have made that my task in the twilight of my life. This includes possible failure, knowing that I make no friends among my colleagues, I'm already on the international list of the dissident scientists who dismantle the official doctrine. If Hindu Buddhist ideas are also processed here, this has nothing to do with religiosity; I came into contact with religions at an age when they could no longer sink into my emotional sense . This is only due to the claim not to be trapped within the boundaries of Western thought.

If Hindu Buddhist ideas are also processed here, this has nothing to do with religiosity; I came into contact with religions at an age when they could no longer sink into my emotional sense . This is only due to the claim not to be trapped within the boundaries of Western thought. During my research, I repeatedly asked myself the question: how can it be that such highly qualified academics can spread such a nonsense as Modern physics? Why should an unlimited universe expand if we can only ascertain expansion in something limited, such as a balloon?

How should a volume curve if only surfaces, i.e. phase boundaries, show structures between two different states, such as between solid and liquid, between solid and gaseous phase, between liquid and gaseous or between gaseous and luminous phase? How can a nothing wave? How can light be a particle? This would require a spatial limitation. Why should a room get a fourth dimension as time? Is past time independent of a route, and does time have a direction? These are all questions that a person with a sound mind poses. The physics is sick! But what is the cause of this world-spanning disease that has infested 20th century physics? »*I did not recognize my physics when the mathematicians attacked it*,« Einstein is reported to have said.

Is mathematics to blame for the fact that physics has become ill? The basis of all mathematics is binary algebra, also known as logic. Logic is the basis of all computers on our planet. All science and technology works logically, only Modern physics beyond our sensory perception? You could suspect a conspiracy behind it. There are indications:

1. I refer to the 2007 edited book *Gottes Universum* by Oven Gingerich [1.02], which is a devotee of creationism and an opponent of the Darwinian doctrine. Gingerich professes a stable belief in a personal, almighty God and finds that science and religion are not only compatible, but complement each other; and he also believes that the way scientists express their view of God and religion requires more balance, accuracy, and moderation.
2. Atheism, with its superhuman idols in the world, spread very strongly in the early 20th century. "Lenin with us" I saw on banners in the former Soviet Union. I only knew the phrase "God with us" before. For this purpose, a counter-movement had to be created, for the miracle of the Immaculate Conception of the Virgin Mary already had its day. On top of that, woman's rights activists also can regard this miracle as a physical abuse. The theory of the Big Bang goes back to George Lemaître, who is based on Einstein's theories. This theory was declared by Pope

Pius XII in November 1951 the scientific foundation of the faith and thus the doctrine of all Catholic schools.

3. St. Hawking, in his book *A Brief History of Time*, reports on a conference on cosmology organized by the Jesuits in 1981, where he presented his idea of a universe that has neither borders nor a beginning or an end. At the end of the conference, the participants in the meeting were invited to an audience with Pope John II, who explained that it was okay to study the evolution of the universe, but they should not investigate the Big Bang, that is the creation of God. So the standard physics begins 10^{-43} seconds after the Big Bang.

Now, as you know, there are many different ideas of God in the world, and that's why people bashed in each others their heads for centuries. Faith does not seem to me to be a good foundation for science. For faith begins where knowledge ceases. When faith embraces science, that is its downfall. Enlightened people since the Renaissance in the West, have pushed back the faith in favor of knowledge in a hard struggle with the Catholic, but also reformed churches. Do we want to put it all at risk by surrendering ourselves to our fate and keeping our distance from logic and mathematics just because they want us to believe, we wouldn't understand anything about it?

My theses:

1. The sciences in the past were divided into basic sciences and applied sciences. The engineering sciences, which are among the applied sciences, have developed their own fundamentals over the last century. Since the emergence of space science, theoretical physics, which is regarded as fundamental science, lost importance to society. It needed new fields of activity and in the second half of the last century it established itself as a pillar of religion. So there are two paths to Theoretical Physics, either

reforming and returning to scientific principles, or disappearing into meaninglessness as it degenerates into a justification of belief.

2. Science in the service of the world view has to do with philosophy. Philosophy, however, is not free of religion; and religion is again a broken reflection of the society in which it originated. Thus you find in the roots of the old polytheistic religions a piece of natural philosophy. The divine principle of the Trinity of Catholicism can already be found in Hinduism. But there the gods embody mass, movement and energy and are themselves expressions of nature around and within us. The feedback in the causal cycle of nature, this original principle, already formulated in Sanskrit, the language of the Indy-European culture, has been forgotten in Western physics. The engineering sciences have rediscovered it. But this ancient philosophy of nature forms the foundation of physics and engineering sciences, namely the theory of the motion of matter in its cycles.
3. A sign of the retrogressiveness of a world view is the inclusion of mysteries and their formulation in a language that the masses of people do no t understand, in order to emphasize the special status of the ruling elites. In addition, a lack of alternatives of the presented ideas is preached. This is intended to cement the special position of this group. In the Middle Ages this was the Latin language and today they use the language of mathematics. But mathematics does not explain the mysteries of Modern physics, because mathematics builds on logic. S o mathematics is abused today, as the holy virgin once was.
4. The past of a development process, including that of the cosmos, can not be determined from the analysis of its current state. That's the insight from system theory. Laplacian determinism is an illusion known at least since the 2nd Theorem of Thermodynamics. The initial conditions of the development of the universe are not knowable and speculation about it does not belong to a serious science. They are constituents of the faith.

5. At the beginning of the twentieth century, the Christian faith was threatened by the spread of communism and the propagation of an atheistic scientific world-view. The new gods were Karl Marx, Friedrich Engels and Vladimir Ilyich Lenin. A counterbalance to this movement had to be found. Although Einstein's fame was initially promoted in Zionist circles, the success of the work of Albert Einstein is seen in the first context. His ideas became the basis of George Lemaître's Big Bang model, which Pope Pius XII declared in November 1951 the scientific basis of the faith. Velikovsky, who also had a Zionist background, was condemned because he did not fit with his *Worlds In Collision* in the concept of a christian world order. Even in communist circles, one tried later to exploit the popularity of Einstein after initial resentment.
6. There are two descriptions for energy: E = hn for the waves and $E = m \cdot c^2$ for the particles. Light quanta have no mass. However, this means that the particle description for light is not correct, since light quanta in the particle image can not have energy. Quantities of effect are neither particles nor waves, but effects. No one comes up with the idea of describing hammer blows as particles, nevertheless, each stroke contains a quantum of effect that hits a nail a little deeper into a board.
7. »*It is well known that Maxwell's electrodynamics-as it is usually understood-leads to asymmetries in its application to moving bodies, which do not appear to adhere to the phenomena.*« This is how Albert Einstein's essay with serious consequences begins, named *On The Electrodynamics Of Moving Bodies.* Nobody has considered in 1905 the consequences of this sentence. Although this sentence seems to suggest that others have also dealt with the subject of symmetrization, such as Hendrik A. Lorentz for example, the work contains no references. Symmetrizing means that the limit for *v* versus *c* of div **E**

approaches zero due to the Lorentz transformation, which is a projective transformation because it contains *t* and *v* that are functionally linked, because the observer would see the world from the perspective of a light quantum. He would see less and less because of the spectral redshift of the light. because the quantum of light would not reach his eyes. In addition, no observer laden with mass can not move at the speed of light because of the mass gain $\gamma = 1 / \sqrt{(1 - v^2/c^2)}$ which approaches infinity for $v \rightarrow c$.

8. The general theory of relativity claims that there is a curvature of space caused by gravity. The space is a mathematical concept. Curvatures can only be observed on surfaces. Surfaces are described by functions. Surfaces separate spaces into subspaces by dividing a space in the outside and inside when the surfaces are closed like a sphere. If it is an open area, such as a plane, the space is divided into two half-spaces. Physical forces work on masses and only on masses, because masses are surrounded by force fields, as Newton has already stated. But forces can not be separated from the masses, as Leibniz, in contrast to Newton's theories, stated in his monadology, whose metaphysical character Kant corrected at the end of the eighteenth century.
9. One can't explain away the dysfunction of 20th century physics with invoking mental illness of certain people. Raphael Haumann [1.03] points to the Asperger syndrome, that he has diagnosed with Einstein and which also occurs frequently in mathematicians. That would remove society from responsibility. Einstein was installed as a monument, comparable to Lenin, to save the Christian faith and this does not happen in scientific controversy. This installation as a monument was only reinforced by hi s social incompetence. Since then, much of the intellect has been sacrificed on the altar of relativity in Modern physics, such as the alleged discovery of the accelerated expansion of the universe, the discovery of the "god particle of Mr. Higgs," the black holes of Stephen Hawking, or the discovery of

gravitational waves. Just as no one in the East was allowed to discuss the theses of scientific socialism, today, in the Catholic public sphere, any criticism of the church-supported Big Bang model of the universe is suppressed, which is supposed to support all of the enumerated "discoveries", And all these discoveries are intended to support the correctness of Einstein's statements. The difference is that the democratic constitution of the Federal German Republic protects the critics of this type of science from persecution. For how much longer?

10. These three works by Einstein caused the dam burst for an unprecedented flooding of the physics with fantasies that contradict all rules of respectable science. Mathematics and theoretical physics w ere no longer used as tools in dialogue with nature but as a justification of belief. Today, the customer is supplied with the physics of what pleases. The customer wants to be entertained, so the Star Trek story is better than worrying about whether the math was applied correctly.
11. When I read reviews on the theory of relativity or quantum mechanics, the criticism usually came from non-specialists and was often even worthy of criticism, such as those of Bill Gaede [1.04] or Raphael Haumann, because often wrong or superficial. The conceptual mistakes remain hidden from most people, though they are banal and perhaps so inconspicuous in their banality. Very few insiders even dare to criticize their subject and, when they do, they have tremendous difficulties in doing their job. The academic representatives of physics are obviously neither able nor willing to break away from the embrace of the Christian faith and to carry out a reform of physics. They have proved that in the past. A reform can only be done from the outside, most of all I see the engineering sciences in a position to do so. However, the public can make their contribution by demanding logically clean explanations from the physicists.

As a taxpayer, everyone in Germany has formally since Jun. 25th 2015 the right to submit requests to the scientific services of the Bundestag in accordance with the Freedom of Information Act.

12. In the cosmos, the electromotive forces dominate over gravitation. All luminous mass is there in the plasma state. Just as the charge separation on earth works on phase transitions, a charge separation also occurs in the cosmos, which is still contested by the mainstream physicists. Huge networks of Birkeland currents over billions of light-years connect stars and galaxies and cause the movement of cosmic objects in vortices. The consideration of physics as closed systems is a thing of the past. If you want to understand the cosmos, you have to model it as an open system of mass-and-energy, far from the thermal equilibrium.

Anyone who wants to set out on the path to knowledge should be given this book as a companion. As in Luther's time, an individual can not handle such a reform, this requires the power of the whole society. The time is not yet ripe for this, but the book can make a contribution to opening the eyes of the reader to an unbiased view of physics as it was cultivated before Einstein and Heisenberg and thus solving the puzzles in space exploration. If that succeeds, it has served its purpose.

Mathias Hüfner

Jena in the 501st year after
Luther's theses against the Catholic Church

2 Something Philosophy Pleasing

»Ignorance of the interconnectedness of all things leads to false perceptions and wrongdoing« - from the Four Noble Truths of Buddhism

When you enter the German search term *Physik und Philosophie* at Google, you get essentially one name, Werner Heisenberg, who wrote a small volume in connection with quantum mechanics, in which he rejected the causal law [2.01].

In April 2012, Ross Andersen, editor of The Atlantic magazine, conducted an interview with the theoretical physicist, cosmologist, and best-selling author of *The Physics of Star Trek* Lawrence Krauss, entitled *Did Physics Make Philosophy and Religion Necessary*? appeared in the journal The Atlantic. Krauss' answer to this question outraged philosophers, because he said, »*Philosophy was once a subject area with content*,« and later added,

> *»Philosophy is a field that unfortunately reminds me of that old Woody Allen joke: 'Those who can't do, teach. And those who can't teach, teach gym ' And the worst part of philosophy is the philosophy of science; the only people I know to read the essays of philosophers of science are other philosophers of science, they have no influence whatsoever on physics, and I doubt that other philosophers read it, because they are quite subject-specific. Understanding what justifies them is therefore really difficult. And so I would say that this tension occurs because philosophers feel threatened - and they have every right to do so, because science makes progress and philosophy does not.«* [2.02]

This quote characterizes the situation:

Physicists and philosophers do not understand each other anymore.

Is that bad? - Yes, it is!

In every theory, there are axioms, sentences, basic statements that can not be proved, that are derived purely from the experience of man. Theologians would say they derive from faith. How strong the influence of the faith is on the natural sciences, becomes aware only when one has come into contact with other cultures and dealt with them without prejudice. Hilton Ratcliffe has made an interesting contribution to this with his book entitled *Stephen Hawking Smoked My Socks - How to Beware Our Reviews: an Astrophysicist's Perspective*. These empirical propositions or axioms, which are based on ancient human knowledge and are universally valid, can be found in the philosophy of ancient cultures.

Philosophy derives from the Greek word *philosophia* and means 'love for wisdom'. For a long time it was the discipline that held knowledge together because it was not limited to a particular field or methodology. This could serve as orientation for more specialized fields. In modern times, philosophy split into more and more directions, making it more and more confusing. Nevertheless, there is one criterion on which philosophies can be divorced. It is the question: **Does the mind dominate nature or does nature dominate the mind?** If the mind dominates nature, is it my mind or a general supernatural spirit? If the question is answered that nature dominates the mind, we are in the natural philosophy, in realism or materialism. But realism does not separate the reality of thought from the reality of the outer world, which ultimately leads to a subjective idealism. The concept of materialism is often avoided because it is ideologically burdened by the Western idealists, who grant the mind the control over nature. Paul Marmet described it by the term physical realism [2.03]. Ultimately, however, we can not avoid the concept of matter, since physics, as the basis of natural science, deals with the laws of motion of matter. Physics has neglected nature's philosophy in recent decades. The consequence was that conceptual errors were established in physics. The term nature derives from the Greek word *physis*. Physics simply means science of nature. Aristotle divided knowledge into physics and metaphysics. The deeper meaning was the emergence and growth, that is, the living. At that time, metaphysics was all that came in his books on the pages behind physics. In the Middle Ages then a change of meaning occurred. The world was divided into nature and super-nature. The supernatural was the divine, super-sensible, and thus metaphysics received a new content.

John Needham has pointed out that Occidental thought always wavered between the idea that the world was an automaton and the theology according to which God ruled the world. He called this the peculiar European schizophrenia. [2.04] German Enlightenment philosophy produced the luminous spirit of Immanuel Kant. Even if his work "*The Critique of Pure Reason*" [2.05] comes across a bit cumbersome - he probably meant that this makes a more docile impression - he has found in it fundamental: We need concepts to think and these concepts have to reflect our sensory perception into the language. "Thoughts without content are void; intuitions without conceptions, are blind." Kant sees language as an illustration of nature through our senses in the spirit. Thus, you can classify him in the group of natural philosophers. His sharpest critic was Arthur Schopenhauer, who just saw the light of day when Kant published his criticism. In his major work *Die Welt als Wille und Vorstellung* [2.06], Schopenhauer raised the term will to the metaphysical and coined the term relativity, which to a certain extent has in it the interchange-ability of observer and observed object as mirror images. He created from Indian philosophy, which documents the material world with divine symbols and holds for the individual prayer more gods to choose from, without adequately transferring this philosophy into Christian thought. Thus came out a subjective idealism, which comes close to that of Georges Berkeley, but which is not included in Indian philosophy. Berkeley's main concern with his work *Principles of Human Knowledge* was the refutation of the concept of matter in order to deprive atheism of the ground. This explains Arthur Schopenhauer's success and influence on the faithful as well as Einstein at the beginning of the twentieth century. Part of the criticism of Kant's work concerned the antinomies[2]) of his logic and his naive use of sets. The exact

2 Antinomy of space and time, antinomy of atomism, antinomy of causal determinism and spontaneity and antinomy of necessary being or not

predicate logic and axiomatic set theory was established only a hundred years after Kant, to which the German Georg Cantor acquired great merit.

2.1 The Matter

> A few years ago I was on an educational journey through Rajasthan. While the Bible is here in Germany at the hotel in each bedside table, I found there one of the sacred books of Hinduism, the Bhagavat Gita. Surely I would have put the book away without much attention, if I had not found the following quote from Albert Einstein on the back cover: "*When I read the Bhagavat Gita and reflect on how God created this universe, every thing's seems so superfluous.*" That caught my interest.

Physics as science is an empirical science based on measuring observed natural phenomena. Conversely, that means: What I can not measure can not be a subject of physics. Physics will not be able to give answers about things that are not measurable. These are specific questions of cosmology as well as questions of particle physics that are outside our range of observation and measurement. But there is in this time the field of Modern physics. The boundaries to metaphysics are exceeded here. 'Meta' comes from the Greek and means 'behind'. Originally, the term goes back to Aristotle, whose eight books dealt with physics and which put the philosophical questions behind them, such as:

- Is there a final sense why the world exists at all and that it is just as it is?
- Are there any gods and if so, what can we know about them?
- What is the essence of man?
- Is there such a thing as "spiritual", in particular a fundamental difference between mind and matter, and does the mind determine matter or is the mind dependent on matter?

The latter question divides the philosophers of the Western world since that time into the supporters of the idealism and the supporters of the materialism. The Hindu philosophy says: The gods are matter! You could transform this thesis into a Western pantheism, which is not justified, because the Indian gods are used as symbols, since their interpretation is multilayered and for the private worship other gods are used as for the description of the world. The key concept for physics is *matter*. Yet, as so many do in Modern physics, this term is not clearly defined, which has caused a mental confusion. Physics is the study of the movements of matter. For some people, there is a strange identity of matter and mass. That's why we need to clarify exactly what we mean by that.

The materialist answer is: matter is the objective reality that exists outside and independently of our consciousness and can be detected by our senses and measuring instruments.

Excluded from matter are all supernatural things that spring from our imagination. The mass confront us macroscopically as a universal property of matter. Microscopically, it is an expression of the non-countable amount of elementary particles. The difference between set and mass is that sets are usually countable, but masses are not. Masses are determined by means of a mass equivalent. Consequently, at the macroscopic level, we are talking about the mass of bulk material, and should keep 'particles' in mind. Another property of matter is its energy content and the resultant movement its particles. So we perceive the mass in bulk in their four different density states, solid, liquid, gaseous and luminous with increasing motion, as well as their phase transitions between the different states of motion. While we perceive the solid state as a definite shape body, the shape in the other states is dictated by external conditions. Each elementary particle is assigned a charge of positive or negative charge, which we perceive via the forces between different mass particles. Energy and movement result from this. Modern physics also claims charge-free particles, the neutrinos. We will return to these concepts in the next chapter.

Our surrounding real existing world is material, by which we mean all that was mentioned above, and all this is subject to a cosmic order.

This includes the enumerated mass, as well as the surrounding force field resulting from the charges of their particles, which we feel as gravitational force or electromagnetic force, that causes the movement of the masses in their different phases by means of acceleration. At this point, the reader will feel the contradiction, since he has learned something else, but the reason for this statement I will give in the following chapters.

Matter exists only as a singular, as does universe (the all-embracing). This means that there are no different types of matter and no additional parallel universes. I mention this because in recent years different claims have been made by the specialist literature. Dark matter, if it does not mean the first three aggregate states, and the parallel universe are in the Kantian sense empty concepts, since they do not originate from sensory perception. Surprisingly, we find the strangest ideas in the specialist literature of the Occident about these concepts.

Tharun Chopra has brought Indian philosophy closer to the Europeans in his book "The Sacred Cows" [2.07]. In Indian philosophy, the root of Indy-European culture, matter has the highest value in its divine Trinity. There the Sanatana Dharma describes the eternal cosmic order. The three main gods, Brahman, the All-encompassing; Vishnu, the symbol of structure formation of mass and life and Shiva the symbol of destruction of the structure; as well as their marital goddesses Saravati, Lakshmi and Parvati united to Shakti to the symbol of cosmic energy. The ancient Aryan gods didn't possess any animals for locomotion. According to the Rigveda, they had aircraft's pulled by fast horses that never tired. This older concept of the gods thus emphasized the ability of the gods to move without restriction and at high speed throughout the universe.

The link between the philosophy of the ancient Indians and the Western world is provided by Pythagoras, whose ideas found their way into the Renaissance era. Pythagoras settled in 530 BC from Greece to Croton/ Calabria in Italy and founded a school there. The Pythagorean's believed in transmigration similar to that of Hinduism and the Buddhism that was born at that time. The medieval church of the renaissance period was concerned with the Pythagorean's. An example of this is the garden design of Villa Lante near Viterbo in the Lazio region, which I had the opportunity to visit for myself.

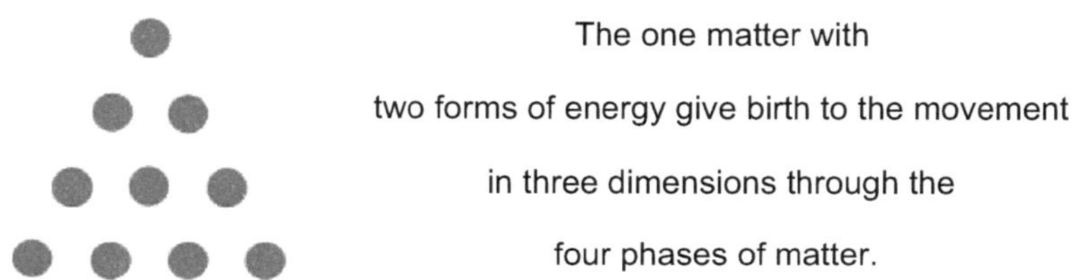

The sum of this we measure in the decal number system

Here the four phases of matter are symbolized. In modern form we can summarize the Pythagorean principle as above seen. From this the Pythagorean's developed a number mysticism that has been handed down to Aristotle to the present day as a mathematician belief. Baruch Spinoza, a philosopher of the 17th century, also created from the ancient sources. While the Pythagorean's have set the spirit over the four phases of matter and chose the pentagram as their symbol, Spinoza saw possibly inspired by Indian philosophy in nature the spirit itself, thus establishing the pantheism which Einstein also adopted as his faith.

The equivalence between energy on the one hand and the product of mass and the square of the speed of light attributed to Einstein describes matter in its most general form. Strangely enough, for him, this equation was never of central importance in his theoretical reflections on the theory of relativity, though he was so impressed by the *Bhagavad Gita*. We will come back to this later. In Indian philosophy, the gods are part of the universe. This cosmic order is unlimited and eternal, symbolized by the wheel of life *Bhavacakra*. Consequently, the existence of infinite and eternal matter is not an invention of modern times, more than a rediscovery. Even the religions of Krishna, Judaism and Christianity are contained in this concept as an incarnation of the eternal Vishnu, the principle of becoming, as an expression of the democratic power of integration of Hindu philosophy. How disturbing are the monotheistic claims of Judaism, from which then developed Christianity and later Islam with their violent missionaries, both of which fought the idea of democracy and as destroyers of other cultures they acquired a dubious fame in history. We believed that democracy was invented by the Greeks. How short-sighted we were? The old polytheistic religions reflect the social picture of a democracy of ruling strata going back much further. The genesis of the Rigveda dates back to the second, pre-Christian millennium. Today we summarize their cosmic idea in the equation $E \approx m \cdot c^2$. The sign $\approx$ means that the energy equals a mass if it were moving at the speed of light, but this is not possible. So this formula represents a non-surmountable energy barrier for elementary particles. Since matter and gods are a unity for Indians, their gods are real. They exist outside the human mind. On the other hand, in Indian philosophy humans live in a perpetual cycle of rebirth in a dream world until they are allowed to rise to Brahman in the real world by dissolving their shape and becoming part of the cosmos.

To this dream world, the subjective idealism of the Occidental style took hold, without the essence of Indian philosophy having been adopted. What remained was only the idea of a subjective, a virtual world like in a mirror without real existing masses. So Arthur Schopenhauer found his concept of relativity. Arthur Schopenhauer anchored this concept in his major work *Die Welt als Wille und Vorstellung*, in which he arranged a will for each object and postulated the relativity between object and observer. This brought him attention, since he subjected his own criticism to the unpardonable *Critique of Pure Reason*– the book was put on the *List of Prohibited Books* by the Vatican.

This philosophical error of relativity was taken over by Albert Einstein without indicating the source. Einstein, as a devotee of Schopenhauer, found this concept much more appealing than that of matter, without considering that the masses of observer and object are usually quite different and it makes a big difference whether the prophet Einstein approaches a mountain or the mountain comes to the prophet. We will see that the concept of relativity goes right to an illustration. Schopenhauer's idea of the relativity between object and observer donated in physics havoc when it took Einstein. Thus, the boundaries between the outside world and its image in mind were blurred by means of sensory perception.

In his *Critique of Pure Reason*, Immanuel Kant introduced two kinds of terms (you could also say Criticisms of Pure Mathematics today), those based on the classification of sensory perceptions, *phenomena*, and those that do not correspond to sensory perception called *noumena*, but emanate from imagination. Today, the term sensory perception should be better replaced by the term engineering perception, as all metro-logical equipment is based on engineering that does not indulge in speculation.

> *»Terms without intuitions are empty, intuitions without terms are blind..«*

and

> *»Thus all human knowledge begins with terms, then from there to concepts, and ends with ideas.«* - *Immanuel Kant*

In order to separate the inside of the mind and the outside of the world, we need terms that separate the mental inner life from the outside world. This is quite difficult, because we have both a picture of our inner world and also the outside world in our mind. There are so-called synonymous terms that can do that if we could agree on when to use the one and when the other term. An example should illustrate this. In physics we should consistently use the term volume in connection with the mass, while the term of space should be reserved for the mathematical concept. A volume of the outside world has basically three dimensions, while the concept of space is not limited in its dimensions.

The metric space, which is equipped with a distance measure and three dimensions, is in principle not limited, because the set of rational numbers is unlimited. We use this space in physics to evaluate a limited volume of a massive body. Likewise, we must distinguish the space from a surface. While the subspace represents a set of spatial points, the objects of space, the surface is the boundary to another subspace. Surfaces are described by functions. In physics, we use the term surface to distinguish different phases of matter. The surface always belongs to the denser medium. Again, a lot of functions may be needed to describe the surface of a physical body. The surfaces then form structures by folding. One of the most important properties of matter is its structural organization. It is undeniable that mind and matter differ in their structure. Masses have a spatial structure and energies have a temporal structure. Our mind extracts information from these structures and assigns concepts to them. It is an imaging process in certain symbol sequences. This is the field of computer science, whose field is mathematics.

Can the mind affect matter? If you follow Schopenhauer, this question is justified. This question is similar to the question: Is the image or the imaged object the reality? Relativists fail on this question. No, the mind can only work on matter through our hands. Who pretends to be able to influence matter with his mind, we call a magician or illusionist. Because he is able to deceive our perception. But not only our perception can be deceived, also, and this is much more the case, our mind is subject to difficult-to-erase errors.

2.2 Mistaken Philosophical Concepts in Modern Physics

In Indian philosophy, you have to make a clear distinction between Hindu and Muslim philosophy. The Muslim influences on Indian philosophy began with the invasion of Mughal rulers in northern India at the beginning of the 16th century under Din Muhammad Babur. They banned all human symbols, destroying much of Hindu culture and used symmetry as their ideals, which was perfectly expressed in the gardens, as replicas of Paradise and in the mausoleum of the Mumtaz Mahal in Agra, the Taj Mahal, which was built by Great Mughal Shah Jahan from 1631 in 17-year construction as a symbol of his love. He is said to have had 72 virgins in his harem, for which young Muslims today as terrorists blow themselves up.

Indian philosophy found its way into the German consciousness only in the 19th century and thus the name Schopenhauer is closely connected, while the rediscovery of the Greek philosophers in the time of the Renaissance fell. Thus our philosophy today may be much more influenced by the Greeks, such as Empedocles, Socrates, Plato and Aristotle, and with it classical physics. Modern physics is based on Schopenhauer's criticism of Kant and Berkeleys subjective idealistic ideas. The weak point of Kant's *Critique of Pure Reason* concerned his ideas about the sets. A logically closed set theory did not appear until the first half of the 20th century, and the beginnings of Modern physics fall into a period of basic crisis in mathematics.

Physics as a measuring science can not do without mathematics. Aristotle treats his philosophy of mathematics in the books XIII and XIV of metaphysics. He criticizes here and in many places Platonism, why he is considered today as one of the first representatives of materialism. In fact, mathematics has nothing to do with physics, it is part of computer science. Physicists mainly use mathematics for the informal description of physical processes. The description is not part of nature, but part of the language and thus a spiritual product. In other words, we reflect the objective reality of matter through symbols in our consciousness or subjective mind, similar to a movie. We have a number of encryption options available. This is the field of computer science, which has skyrocketed in the last 60 years and found insufficient access to physics.

The physics of the 20th century has influenced Albert Einstein like no other. He is celebrated as the genius of the century. Was he really it? Many monuments of the last century have been dismantled.

»When we work on something, we descend from the high logical steed and sniff around on the ground with our noses. After that, we will blur our traces to increase God's likeness.«- *Albert Einstein*

He borrowed two of his basic ideas from Schopenhauer, whose admirer he was - symmetry and relativity. So it is not surprising that two fundamental errors have crept into the philosophical foundations of natural philosophy of the 20th century.

Another mistake was made by mathematics itself in physics. These fundamental errors together shaped the development of Modern physics. It was favored by an increasing specialization and a lack of superior ordering discipline, which was the nature philosophy once.

2.2.1 Do you believe, Nature is Symmetrical and Relative?

The idea of symmetry and relativity Einstein brought with his paper *On the Electrodynamics of Moving Bodies* [2.08] of 1906 in physics. In particular, the symmetry in the cosmos would be predominant because it should be an expression of the divine order. It is the idea of the divine order in Islam. Symmetry means balance, stasis and death. Life is based on self-similarity, dynamic balance and self-organization. I still remember the saying of an Islamist fundamentalist: "*You love life, we love death*" [2.09] Since Einstein's work in 1906, theoreticians have been trying to symmetrize physical processes and substantiate this with aesthetics. This is most surprising, since in Western art the golden section and the Fibonacci sequence have been regarded as the measure of beauty since the Renaissance. [2.10] [2.11] and the symmetry in the visual arts is frowned upon. The golden section can be traced back to the Pythagorean's, as it can be deduced from their pentagram. Each line of the pentagram is cut approximated by another of these lines in the ratio of an irrational number 1.6180 ... .

In my youth I was a member of the Lindenau painting school in Altenburg, Thuringia Germany, where I learned the image composition. But then I decided in favor of physics, because at that time I thought that it was not as ideologically charged as the visual arts. I recognized this error quite late.

We noted above that matter has both a temporal and a spatial structure. Structures can be chaotic, self-similar, and symmetric, requiring an ever stricter order in all dimensions in this order. So there is only one point symmetry and that is the ball symmetry. All other symmetries refer to surfaces along a straight line. By relating this structure sequence to the wheel of life, chaos follows symmetry, as the second law of classical thermodynamics suggests. A less strict order requires self-similarity, and without any order, chaos comes along. Thus, theories of symmetry and super-symmetry are quite specific and are unlikely to provide useful, generalized results.

The situation is different with the similarity and self-similarity that you find at the boundary layers of two physical phases. The fractals invented by Mandelbrot [2.12], with their broken dimensions, have shown considerable success in reproducing natural structures. One of the preferred methods in research is the classification of similar objects. We know this from all sciences. To form terms, we need the classification. The mathematical concept is the set theory. A set (or class) consists of mutually similar objects. The biologist speaks of the class of vertebrates. The chemist speaks of the amount of chemical elements. The physicist, on the other hand, speaks of the mass without specifying this mass more closely. For him it is only important in which energetic state the mass is. At every level, the game repeats itself. We have objects and summarize them into sets. Symmetries are very specific similarities that are limited to surfaces. The simplest symmetry is the ball symmetry. A second form is the mirror symmetry along an axis of symmetry on a surface. Crystals tend to form symmetrical shapes out of a liquid. A third shape is the rotational symmetry about a center. The latter two symmetries are ob-

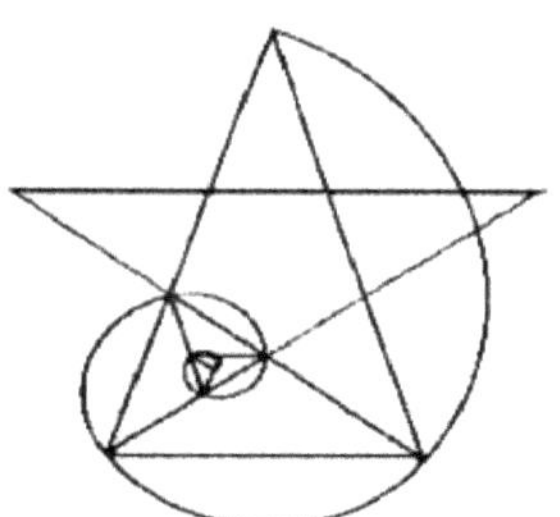

Figure 2.1: Fibonacci spiral and pentagram

served in biology. But looking closely, we find that these symmetries are usually disturbed in nature. Any detail breaks the symmetry. If that were not the case, you could not distinguish natural objects. It is therefore better to speak of self-similarity.

From the symmetry Einstein concluded the interchange-ability between object and observer. Schopenhauer and Einstein believed that one could easily exchange observers and observed objects. Einstein invented the concept of inertial system for it and explained it that way: It does not matter physically whether the train is moving and the platform is resting or the train is resting and the platform is moving. From the viewpoint of the energy balance you can not agree and that is decisive for the physics. The term inertial system only pretends to be physical. It is only the observer's coordinate system.

2.2.2 Mathematics would not be a Concept, but Objective Reality

A position among mathematicians also represented by Einstein's friend Kurt Godel, is that mathematical objects (numbers and geometric figures) and laws are not concepts that emerge in the mind of the mathematician, but are given an existence independent of human thought. Mathematics would therefore not be invented, but discovered. By this conception the objective, thus interpersonal character of mathematics would be met. From this it is deduced that one must find a fictitious mathematical object in nature.

As significant as Godel's mathematical accomplishments are, so devious is his philosophy. In his late work, Godel undertook in 1941 a reconstruction of the ontological proof of God by means of modal logic. In contrast to the two-valued statement logic, the modal logic is a quadrivalent logic which besides true and false also makes the values possible and necessary. [2.13]

Godel's concern was to provide evidence that an ontological proof of God could be conducted in a way that would meet modern logical standards. A version of this proof should now have been checked for correctness by machine-aided procedures. The only evidence of existence beyond our consciousness is the collective sensory experience. Therefore one only repeats that one has learned that this thing exists. Furthermore, the definition of the perfect being according to Kant presupposes its existence. The ontological proof is therefore simply a circular conclusion or a tautology, no matter what logic you use.

Mathematics is a language and not a natural phenomenon. As such, it is not discovered, but developed inter-personally through definitions and symbols. The objects of mathematics are the numbers and the number systems along with the associated operations. But counting is a human activity. We do not find numbers in nature, but sets of real existing objects to which we attach labels in the form of symbols or words. Different number systems are common. The most common number system is the decimal system, but even the sexagesimal system is still in use in the circle graduation In succession the development of computers, the binary system has prevailed everywhere, without us being fully aware of it. The binary system is based on the two switching states 'current flows' and 'current stops flow', to which we can attach the labels 'true' and 'false' or 1 and 0. Without having to prove anything, everyone can convince themselves that the laptop or the smart-phone can perform the most difficult mathematical operations today. The magic word is called algorithm here. An algorithm is a calculation rule, a set of verbal instructions on how the computer must perform the operations. While in the early days you had to translate this algorithm into a binary sequence of numbers for the computer, today there are corresponding translation algorithms that take this work away from humans. However, a job is not taken from you, that is the strictly logical thinking. Today there is a widespread fear of artificial intelligence of computers. In any case, this intelligence requires algorithms that an intelligent person has come up with. A computer tolerates no logical inconsistency, you realize that at the latest if you want to get a written algorithm running on a real machine. Today, while you learn to press keys, use your computer mouse, and swipe across the screen, you call it computer science, but before mathematics there is a peculiar awe in most people, which eventually leads to fear. Of course that leaves a lot of room for myths. Nothing is more logical than mathematics and its basis is the bivalent algebra. This is the basis of the number systems and the basic calculations. All higher types of arithmetic can be attributed to basic arithmetic using algorithms and these in turn to the two-valued logic "true" and "false". In between, there are no more secrets. This is unbearable for most people because the machine has no sympathy for human "logical weaknesses". If you rely on mathematics, you can not talk out that something illogical is only to be understood on a higher level, so you would need a higher level of qualification. The logic of humans is usually disturbed by hormones in the brain. It's obvious that the hormonal system is older and it works faster than the logical system. These control the emotions, including the religious ones. You can not condition logic, but feelings.

The fact that you can generalize mathematical concepts arbitrarily, but that nature is limited in the choice of repetitions, opens a door for metaphysics. This limitation is the framework for the deductive logic that Karl Popper studied. [2.14]

2.3 Deductive and Inductive Logic in Research

Logical inconsistencies besides misinterpretations are the most common mistakes in the research work. There are also inductive conclusions. Karl Popper also dealt with the question of inductive reasoning and wanted to avoid this conclusion as far as possible. In order to gain insights, they are usually unavoidable. As mentioned above, logic has something to do with an algebraic system based on 2. This system is also called Boolean algebra, named after George Boole, since it dates back to his logic calculus of 1847. [2.15]

2.3.1. What is Truth?

First, we have to answer the question, What is truth? Truth is the evaluation of a statement with the value 'true'. This is nothing more than assigning a numeric value to a variable, except that in this case I have the natural numbers available for the evaluation, while the logic only knows the values true and false. Because truth is a rating, it is so controversial. The evaluation of statements is interest-related. Those who are in search of the truth should keep this in mind. The stronger the interests, the less likely it is that a controversial statement is true.

The philosophers alone distinguish 8 theories for their evaluation. The most original is the correspondence theory, which goes back to Aristotle, according to which there must be a correspondence between thinking and reality. Marxism adds the idea of imaging between thinking and reality. In the next step, the logical structure of the sentence must agree with the structure of the fact that it represents. Finally, in the correspondence theory the consistency of a deduced statement to the system of accepted statements must exist. Jürgen Habermas, on the other hand, advocates a consensus that is brought about in a discourse in an ideal speech situation. Since there can be no ideal speech situations, the truth is subordinated to power interests. Thus Pope Pius X claims in his encyclical *Pascendi Dominici gregis* a Catholic truth for himself [2:16].

For the natural sciences and technical sciences as empirical sciences, practice (eg, experiment) as practical proof is the primary and sufficient criterion of truth. Both sciences, like the truth itself, have an objective character and are not negotiable. So far the theory. The more complex and more expensive an experiment becomes, the more difficult it becomes psychologically to respect the failure of an experiment as the truth, and the more difficult it becomes to find independent judges on this issue. In this situation are today's researchers. Today you can no longer do physics without the help of engineering. Here, a conflict of interests between engineers and physicists emerges. The goal of the engineers is to get out something useful to society, while physicists settle for an idea that is as hard to refute as possible. This brings us to the basic problem of cognitive logic.

2.3.2 The Basic Problem of the Knowledge Logic

Every empirical science uses the inductive conclusion by generalizing a special observation. If, for example, you have observed enough times that the sun is at noon in the south, then you conclude that it always does. However, if you start your journey to the south, you will find that this statement is suddenly wrong. South of the equator, the sun is at noon in the north. However, deducing deductively from the generality of a particular fact saves one from the surprise that the statement can be wrong.

Karl Popper, who has studied the logic of research very intensively, now attempts to circumvent the fundamental problem of the logic of knowledge already preoccupied by Hume and Kant by avoiding inductive conclusions, which he ultimately can not succeed in doing.

- That's the problem of induction: special sentences are generalized. Such a conclusion can prove to be wrong. You can also formulate the induction problem as the question of the validity of general empirical propositions, of empirical-scientific hypotheses and theory systems. (Popper) You must ask yourself when the induction sentence is permissible and when not. A very clear example of a false induction closure is given by the following example:
    1. No cat has two tails. A cat has a tail more than no cat. Induction conclusion: Cats have three tails. In this simple example, the first sentence contains a negation. The second sentence combines this negation with a positive statement. This is obviously not allowed in the induction.
    2. Another example of induction is: We have a set of measurements and approximate the measurements by a polynomial. We can either guess the polynomial from the curve or construct it according to the method of statistical experimental design, a method widely used in the art. Then, within the

measured interval, we can almost certainly make true predictions about the expected values in this interval. But the farther we move away from the studied range, the more unreliable the value calculated from the polynomial will be. Physical laws without validity data have this uncertainty. These are probabilistic statements that most people are not even aware of. The induction problem seems insurmountable.

- This results in the problem of demarcation: From the two examples, it immediately becomes clear that you must distinguish the scope of inductive true statements from the range of their false statements. Both examples have completely different demarcation criteria. From this it can be deduced that the inductive method of reasoning will not have a general criterion of demarcation between its admissibility and its inadmissibility. To put it more generally, for empirical science there is no universally valid criterion of demarcation from mathematics or, for example, metaphysical systems and those of the imagination. Popper therefore generally rejected induction logic. He says: »*The induction-logical demarcation criterion does not lead to a demarcation, but to an equating of the scientific and metaphysical theory systems, not to an elimination, but to a collapse of metaphysics in the empirical science*« [2.14 p 9] Is then no more generalization permissible? That would be fatal. But the danger of a fallacy persists.
- If we grasp this reasoning logically, then we can distinguish two demands that we must make of the empirical system of theories: It must represent a self-consistent, possible world and must satisfy a demarcation criterion, i.e. it mustn't be metaphysical (it must represent a possible world of experience). This demarcation criterion is known a priori, since it springs from experience and thus from sensory perception, but not in general.

That makes things so problematic. A hypothesis can usually not be falsified after their preparation because the knowledge is insufficient. Popper therefore set up two methodological rules, according to which the review of the scientific sentences must be made.

1. The game 'science' has basically no end: Who one day decides not to continue to review the scientific sentences, but considered as about finally verified, who is out from the game of science. (like the supporters of standard models, particle physics and cosmology)
2. Once established and proven hypotheses must not be dropped "without reason"; Among other things, "reasons" are: replacement with other, more verifiable hypothesis; means Falsification of the implications.

In addition to the two rules of Popper, there are still some symptoms that point to a pathological science and metaphysics in physics. These were compiled by Irving Langmuir in 1953 at Knoll's Atomic Power Laboratory (KAPL) in a lecture and provide orientation. [2.17]

- The maximum observable effect is caused by a cause of barely observable intensity; the size of the effect is generally independent of the size of the cause.
- The effect has an order of magnitude which is at the limit of observability; Due to the low statistical significance of the results, many measurements are necessary.
- There is a claim to very high experimental accuracy.
- Fantastic theories, which often contradict the experience, are set up.
- Criticism is reciprocated with auxiliary assumptions. Example of an auxiliary assumption: Max Planck justified with the auxiliary assumption that the physical effect can occur only in multiples of an action quantum, the quantum physics.

- The ratio of followers to critics increases initially, then gradually back to zero.

The latter two criteria are no longer valid. The peer-review system, which aims to ensure the quality of scientific publications, is supported by the mainstream of the community and is thus not impartial. Critics are suppressed and followers are favored over critics.

2.4 The 4 Basic Axioms for Physics

If you want to create a theoretical basis for a science, you need a set of basic concepts with their exact definition and a set of true sentences that are valid without contradiction. These sentences can not be derived from the discipline. They are universal and should therefore be derived from philosophy. That's easier said than done. We have seen that the symmetry principle does not belong. They must be principles that are old enough and repeated in many philosophies.

The conservation of matter: The Indian gods represent matter, as we have stated above, but no philosophy can make a valid statement concerning the beginning of the world. These natural gods are considered immortal. It follows that matter is unlimited in time. Matter can not arise or pass away. It changes in a constant cycle of origins and transgressions, symbolized by the gods Vishnu and Shiva, symbols of **the unity of opposites** and the source of the forces that turn the wheel of destiny, the *dharmachakra*. You find this basic principle also in the *yin* and *yang* of ancient Chinese philosophy.

The causal principle of cause and effect of becoming and passing away is therefore in Indian philosophy not focused on an absolute beginning, as in Western philosophy, although such thoughts are not foreign to this philosophy. While in the occidental world the causal concept always implies causal chains with beginning and end, the thought of recursion is prevalent in Far Eastern philosophy. Here it is contained in the principle of rebirth, where the previous life affects the subsequent life. The causal principle is fundamental to mathematics. From this follows immediately the concept of function and equation, from which the independent variable results as cause and the dependent variable as effect. In the engineering sciences you find the principles of the control, the foreign and self-organization, whereby the self-organization always contains the element of the feedback. Self-organization is therefore described using coupled differential equations, as we shall see later.

Symmetry became a dominant principle in physics at the beginning of the last century. Islam, with its ban on a pictorial representation of life forms, has chosen symmetry as its beauty criterion. The Occidental Beauty Symbol Criterion is the Golden Ratio. In fine art, symmetry is boring. But beauty symbols have nothing to do with physics. **Self-similarity** occurs on different scales instead of symmetry limited to surfaces; self-similarity also includes symmetry as a mirror similarity. Similarity and opposites are distinguished by the fact that similarity refers to structures and that the unity of opposites creates an unstable equilibrium through their force action, becoming unstable in the sense that it constantly balances itself in a certain direction. These axioms are discussed in section 3.3. again treated in more detail.

2.5 Space and Time

Einstein has tried to give space and time a physical meaning. He overlooked the difference between volume and space, as well as the difference between clocked energy flow and time. While space and time are mathematical concepts, in volume and energy flow we are dealing with real existing things. Einstein's thinking was trapped in the idea of the interchange-ability of observer and observed object, of the identity of object and its image. Nothing else means its relativity. Both Heisenberg and Einstein included in physics the observer, which was a revolutionary act. But both have forgotten the laws of image formation between mental image and objective reality.

2.5.1 What is a Space?

It is interesting what Kant says in his *Critique of Pure Reason* for Space:

> *»Space does not present any quality of any thing in itself, or in its relation to each other. It is only the subjective condition of sensuality under which external intuition is possible. Accordingly, we can only of personal view point speak about the space.«*

Here there is no clear relation to interpersonal objectivity. The camera was not invented yet, but the relationship between man and the environment is already indicated. All the bodies around us have a volume, they have an extension in every direction, they have a mass, and they exert forces on each other causing movement. Mass and force as well as the relation of mass and volume, the density, are physical concepts. What then is a space? – Stand upright. Then you can see what is in front of you and behind you. You can specify what is to the right of you and what is to the left of you. You can also tell what is above and below you. This creates an order in your mind about your environment. You have probably already noticed that with the 6 relations you have described three independent directions, each with opposite orientation. That was a process that took place in your consciousness and that did not trigger any effect or movement in your environment, so Kant found that this was subjective sensuality. So space has nothing to do with physics but with the reflection of the outside world on our imagination. But that already knew Euclid. Therefore space is a mathematical concept, in contrast to the concept of volume, which is filled with mass and forces.

In order to understand the concept of space, we must borrow for some basic concepts of mathematics that I can not presuppose. First, we have the mathematical terms *element* and *set*. We distinguish element-symbols of set-symbols by small and large letters. Any objects, such as numbers, letters, symbols, pictures, or things of daily life, we call elements that can be summarized into a set under at least one particular common property. For example, taking the set of rational numbers, we can represent that set as a number-ray.

If the elements of the set itself are sets and there are ordered relationships between them, we call such a total set *system*. We have encountered the concept of system several times in this book and will encounter it more often in the following. A closed system consists of a lot of inmates and a lot of limiters that the inmates don't want to let out. Closed systems have fixed impermeable boundaries and the boundary belongs to the system. On the other hand, there are also open systems where there are border crossings or no borders at all. They have permeable boundaries to their environment with which they are in exchange. The boundary of a set is also called its edge. An infinite set has no edge.

For example, a point set is a system because one point contains elements of multiple sets. We are accustomed to represent points pi (x, y, z) as elements of the endless point set P, as coordinates by three real numbers. We call this *open system* space, if it is guaranteed that the three subsets, called width X, height Y and depth Z, are independent of each other. This means that the respective cut sets that can be formed among each other are empty. They contain no elements. Any, who needs further information on set operations should be advised to read *Einführung in mathematische Methoden der Kybernetik* by Wilhelm Kämmerer. [2.18] It is easy to see that such a space definition is very general. For example, a color space is given by three primary colors red, green and blue. Each point in the color space then symbolizes a certain color. But this room has not defined a length. The Euclidean space has a distance measure that follows from the generalized theorem of Pythagoras. The set of Euclidean space points is characterized by $R^3 = \{X, Y, Z\}$. In the following, we want to understand under space, if not mentioned separately, always the Euclidean space. So we see that space, unlike volume, is a mental concept. A volume can be described with a Euclidean subspace. Using other spatial concepts violates the reality principle and are useless to describe the us environmental nature.

A space is needed in physics to establish an ordering relationship between the observer and the surrounding objects. It creates order for the observer by assigning it to the real existing volume of a mass to determine distances and sizes among each other. It forms an orientation framework. The concept of space, as used in mathematics today, is a relation of order between real objects and a measure. However, this order relation is much more complicated than such relations as above, right or forward. For a metric space, you first need independent numerical sets of characteristics for description, in our case for the preferred directions height from above and below, length from front and back, and width from right and left. The orientation provides the sign. For example, the upward direction is replaced by the unit vector z. We declare vectors with small bolt letters. Then the direction down is indicated by -z. The unit vectors x and y stand for the other two preferred direction pairs. The spatial point {0,0,0} is always the point from which the observation takes place.

For the feature sets to be independent of each other, these features must be perpendicular to each other in pairs. Only then is it possible to change a characteristic without any other feature being affected by this change. This is expressed by the scalar product of the unit vectors as follows:

$$\mathbf{xy} + \mathbf{yz} + \mathbf{xz} = 0 \qquad (2.01)$$

Equation (2.01) is satisfied only if the scalar products of all the unit vectors are individually zero. But that is only if the cosine of the angle, which the respective vectors include, is zero. This is true if and only if it is a right angle. But that's not enough. If you form a triangle in the rectangular frame of these features, these triangles must all have an angle sum of 180°. **For curvilinear coordinates, such as on a curved sur-**

face, the independence is no longer given, since a surface is described by a function in space. This is an overlooked circumstance with serious consequences for physics, as we shall see later.

Since we also need quantitative statements about the distances of our objects in space in order to measure something, we must agree on a unit for the distance between two points. Since the French Revolution, the meter has been used in most countries of the world. This unit must be invariant to turns of the coordinate system that spans the space. To ensure this, the Pythagorean theorem is used to determine the distance by means of preferred directions, and this measure is then called a metric. The distance measure for the space is then expressed by the preferred directions and their values, the coordinates:

$$\Delta s = \sqrt{x^2+y^2+z^2} \tag{2.02}$$

We call spaces that have such a measure of distance as metric spaces in contrast to topological spaces for which there is no measure of distance. For topological spaces, the only general requirement is the functional independence of the features spanning the space. Relational databases are based on this principle. If a point moves in any direction except in a preferred direction, then one always uses a linear combination of several preferred directions to describe this direction. The number of preferred directions is also called the *dimension* of the room. So our observation space has 3 dimensions or 3 degrees of freedom of linear movement. The observer is always at the origin of this frame spanned by the features. In German parlance, the term co-ordinate system has become common. The fact that the number-ray of rational numbers is dense and infinite does not allow space to expand. However, infinity is not a number symbol that you can count on. If they speak today of the ever-faster expansion of space, that is mathematical nonsense, even if it was awarded a Nobel Prize in 2011 [2.19], which did not prevent a peer review system.

But you can also do something different to represent spaces, for example, by opening a room with a distance and two independent angles. By now you should be convinced that the concept of space has something to do with order and not with physics. So again Einstein should have asked Kant, instead of Schopenhauer [2.20], who saw space and time in the object of intuition. Therefore, Einstein tried to give space a physical meaning instead of establishing the relation between intuition and physical reality of the volume of the mass.

Where remains the physics then? Physically, we have mass points (atoms) between which distances exist, depending on how much force between them provides for their cohesion. These masses with their forces and distances fill a volume. This is the meaning, when you assign points of space to mass points. This assignment is called a relation. The difference between space and volume is that a volume has a moving mass of energy, while a space is a mental measure of a volume, which is why volume is the physical equivalent of mathematical space. We have to keep this difference in mind if we want to evaluate theoretical physics.

Masses can expand in volume due to energy input. A space establishes a relationship between our intuition and the masses of the outside world, it brings us an order into the outside world, which is why this relationship between volume and space can be described as an ordering relation.

2.5.2 What is Time?

For Einstein, time was what watches show. An analog clock is a machine that transmits a potential energy via a clock pulse in a rotary motion of two pointers and these angles serve as a measure of time. The people before the invention of the clock then probably had to live without time? Certainly not. How should they determine the exact time of sowing? Let's be very pragmatic. Let's just look into the past! People knew the concept of time even before watches, as we know them today, were invented. When we speak of time, we mean a period of time, since there is no absolute zero of time. But this zero point, like in mathematical space is a matter of definition. An event will never be able to be observed by two different observers at exactly the same time, just as two different observers can not choose exactly the same observation location.

The pulse for our time is the result of the movement of the Earth itself and around the Sun and the associated energy flow from the Sun, which synchronizes all life on Earth. With the rotation of the earth and the course around the sun, humans have coordinated their lives on earth and determined the sowing point, harvesting and even distances. What's more, the sun's energy flow to Earth synchronizes any life on our planet. Even today you can find distance information in hours on old mail columns. Sundials, water clocks, hourglasses, and candles for timekeeping were also used in past. At night, with the sky clear, the movements of the stars and the phases of the moon were important for the timekeeping. As industrialization increased, people's lives became more hectic. The time had to be divided into smaller and smaller units. For small time differences needed more accurate clocks. Only the introduction of mechanical pendulum clocks brought progress in timekeeping. The principle of the clock is usually a clocked energy flow. A weight is brought to a higher potential, which moves down again and this movement is inhibited by means of pendulum and anchor, whereby a clock pulse is generated, which drives a wheel-work. The energy flow is also the drive here, generated by a weight or a spring. Even our digital clocks need a pulsed energy flow in the form of the electric current of a battery and an oscillating circuit. All these clocks we synchronize with the movement of our earth around the sun and its self rotation. Actually, we have a local time for every meridian on earth. From these we derive political time zones.

The old idea of determining distances by means of time was taken up again when the 'Meter' was defined. When an object moves evenly over a period of time, it always travels a certain distance. The ratio of distance and duration of time is known to be the speed of movement. Thus, the 'Meter' was fixed on the basis of the constant vacuum speed of light in relation to the frequency of the red cesium line on the earth.

Time has only so much to do with physics as to establish a relationship between an observer and two observed events of a timed movement. The two observed events mark the beginning and the end of the observed movement, regardless of whether the movement then stops or not. This results in a time and space distance. Once you have a distance, you can relate it to other movements and distances. On this basis, physical measurements are made and order to the world. Time thus proves to be an ordering relation based on the observation and counting of cycled energy flows. The observation is for now a subjective process. The objectivity of time is established by the social agreement on the type of clock and the beginning of the counting of cycles. When choosing the clock, care is taken to avoid disturbing physical influences on the clock. Einstein and Kant describe two different aspects of time without Einstein ever had been pay attention to Kant's statement.

Time is based on counting the cycles of clocked energy flows. It establishes a relationship of counted cycles to the orbit of planet Earth around the Sun and to Earth's rotation. In this sense, we get another ordering relation.

Time measures can therefore be defined optionally, but this does not mean relativity in the sense of Einstein. In the vastness of the cosmos you will find enough stable cyclic movements that would fulfill this requirement if humans ever had to leave our earthly paradise, which would energetically be a problem.

It is hardly conceivable for normally talented people that academics have seriously thought about the symmetry of time and thus wanted to override the causal law, because they didn’t bring the mechanical dynamics and the thermodynamics under a common theory. Even of imaginary time is spoken, which has no existential meaning. While the mechanical equations of motion were symmetric over time, thermodynamics isn't symmetrically because of the second law. This led to a mystification of time and it tuck a long period to realize that time has only one direction. It is sometimes upsetting what a mischief academic education can do for.

> *»I gloomily came to the ironic conclusion that if you take a highly intelligent person and give them the best possible, elite education, then you will most likely wind up with an academic who is completely impervious to reality.« -* *Halton Arp*

2.6 Is Einstein's Space-time truly Four-dimensional?

No scholarly work of the 20th century has influenced the mindset of physicists as much as the work under the harmless title *On the Electrodynamics of Moving Bodies.* »*There are probably only three men who can think in four dimensions,*« headlined the newspapers at that time. One of these three men was Hermann Minkowski, who from 1896 was also allowed to count Einstein among his pupils at the Polytechnic in Zurich. However, the idea of generalizing the concept of space came to him only two years after Einstein's memorable work of 1905. Thus, the special theory of relativity was created without the generalized concept of space. Minkowski relies on the work of Bernhard Riemann of 1854 [2.22], where Riemann gives the following distance measure

$$x_1^2 + x_2^2 + x_3^2 + x_1x_2 + x_1x_3 + x_2x_3 = ds^2 \qquad (2.03)$$

without accounting for what that means. He deduced of a curvature scale from this without noticing that he was moving on a surface with ds^2, so he mistakenly called this curvature *space curvature*. This does not change too, if he multiplies all x_i still with some coefficients. As in (2.01) for the mathematical space does apply,

$$x_1x_2 + x_1x_3 + x_2x_3 = 0 \qquad (2.04)$$

In order to form a space, there the components x_1, x_2, x_3 must be independent of each other, which means that their scalar products are individually equal to zero. With this, Riemann's distance measurement passes into the Euclidean distance measurement of the intuition space.

Einstein first introduced his space-time to his General Theory of Relativity. However, in order to be able to assess the special theory of relativity without getting tangled up in Einstein's own explanations, we must explain the generalized concept of space.

If we want to expand the space by one dimension, we can do it formally without having to consider the reality of the view in our intuition. We define a mathematical space $R^4\{x_1, x_2, x_3, x_4\}$. For these dimensions to be independent of each other, the sum of all pairwise linear combinations of the x_i must give to zero,

$$\boldsymbol{x_1 x_2} + \boldsymbol{x_1 x_3} + \boldsymbol{x_1 x_4} + \boldsymbol{x_2 x_3} + \boldsymbol{x_2 x_4} + \boldsymbol{x_3 x_4} = 0 \qquad (2.05)$$

Equation (2.05) tells us that all four vectors x_i are perpendicular to each other, even if our spatial sense fails. If we replace x_4 or any other unit vector with time t, then time would have to become a vector and be perpendicular to the path. But when the time is perpendicular to the path, there is no speed anymore because the vector division is not defined. Which reality should that be?

The distance in this mathematical space would have to be defined as follows:

$$\boldsymbol{x_1}^2 + \boldsymbol{x_2}^2 + \boldsymbol{x_3}^2 + \boldsymbol{x_4}^2 = d\boldsymbol{s}^2 \qquad (2.06)$$

However, this was not defined by Minkowski, but as in my textbook of Theoretical Physics [2.21], which already aroused my suspicion as a student:

$$\boldsymbol{x_1}^2 + \boldsymbol{x_2}^2 + \boldsymbol{x_3}^2 - c^2t^2 = d\boldsymbol{s}^2 \qquad (2.07)$$

When graphically represented, we obtain, without restriction of generality, Figure 2.1, since *ct* is nothing but the path traveled by a ray of light in R^3 from the source to the observer, or an other rectilinear path. In

Figure 2.1 this is not shown exactly because otherwise the arrows would overlap.

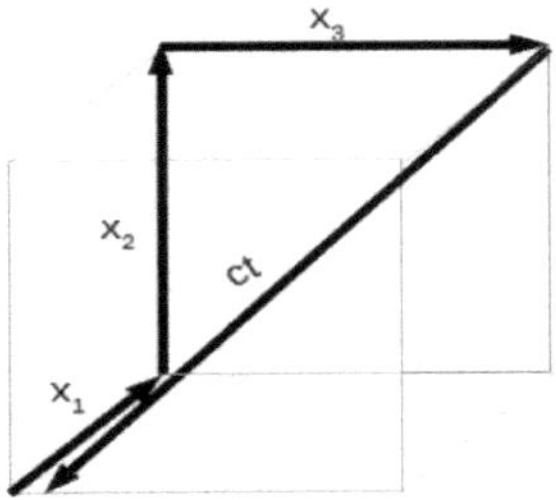

Figure 2.2: To the distance in Einstein's spacetime

So *ds* is zero for the observer. With such a hidden zero you can do the most amazing math magic tricks. In other words, there is no distance measurement. in Einstein's space-time. That had amazed me as a student, but I had not had the courage to criticize that because I could not figure out what the purpose was of this confidence trick. I simply believed in the authority of my teachers.

The space-time is nothing more than a sham, an illusion, because time and speed of light have no direction, so they are not vectors, and because speed is a relationship between path and time, which violates the requirement of independence for the space definition. I need a light source at the arrowhead of x_3 if the thing is to make sense. Then I can set $ds = 0$ because the radius ct corresponds to all directions of observation and is also my distance measure. Look at figure 2.2.

With this simple demonstration it has been shown that the theory of relativity is deprived of its mathematical basis. The assertion that relativity is a generalization of classical physics proves to be a misinterpretation. Generalization means that the newly added element inherits all the properties of the already existing elements. Consequently, time should also have a direction independent of all spatial directions. But that contradicts the definition of speed, and in mathematics there is no quotient of two vectors. No matter what conclusion the theory of relativity draws, as long as this fundamental contradiction remains, all statements of the theory are contestable.

Space-time, contrary to the claim, is not four-dimensional, and the measure of distance is not a measure, since light propagates spherically, which is why ct can take any direction.

The idea of space-time comes from Hermann Minkowski [2.23], Einstein's teacher. He believed in 1907 that he could construct a non-Euclidean space in which the theory of relativity could be formulated. What he didn't understand was, that Non-Euclidean geometry is only applicable to curved surfaces, which are functions, but not available to spaces. In metric spaces Euclidean geometry is valid. Therefore, the cosmos in the Relativity is always presented as an inflated balloon, as if we were two dimensional flattened lice. The volume which is to be compared with the space, however, is the content of the balloon, not its shell. That's the thought trap that most people slip into. We return to this topic under sections 4.7 and 6.1. The world of relativists does not describe our real world.

> *»"The world is my imagination" - this is the truth that holds in relation to every living and knowing being; although man alone can bring them into the reflective abstract consciousness: and he really does this; that is how philosophical prudence came to him. It then becomes clear and certain that he knows no sun and no earth; but only one eye that sees a sun, a hand that feels an earth; that the world which surrounds him is present only as an idea, i. consistently only in relation to an Other, the imaginary which he is self.«*
> *- Arthur Schopenhauer*

Paul Marmet would describe Schopenhauer's philosophy as mental realism as opposed to physical realism. Materialists call this philosophy 'subjective idealism', since here too the primacy of matter is not recognized. It is bad that wrong philosophical concepts and wrong choice of terms can affect science and inhibit its development. Now that we have just discussed the relationship between mind and matter, recognizing that our brain is a material object in which we reflect the world and describe it with language and symbols, we orient toward the observer in the next chapter and examine the imaging process.

Thinking is the hardest task there is. This is probably why so few people are concerned with it. - Henry Ford

3 The Observer

»Stay always attentively in watching your actions and keep here nothing unworthy for your attention.« - *Confucius*

The revolutionary element in physics in the twentieth century was the introduction of the observer into the physical considerations of Einstein and Heisenberg, or as Ilja Prigogine put it, the discovery of new ways of dialog with nature.[3.01] But already in the previous chapter we found that this dialog was still full of misunderstandings. But that's more up to the observer and not the nature. Explaining the zeitgeist, Prigogine writes about the experimental dialog:

> *»The experimental dialog with nature discovered by Modern physics is based less on passive observation than on practical activity, the manipulation of physical reality so that it is as close as possible to the theoretical description.«*

That means that I come up with a theory and stage the experiments in such a way, that they confirm my theory without any account of how my manipulations retroactively influence the observation result. Here science is used as a justification for ideas of an elite.

But this is in direct contrast to what Karl Popper [2.12] said about the research logic, namely that I have to come up with an experiment that can disprove my theory, and if the experiment fails, then there is some chance that my theory is correct. This is because inductive reasoning needs to be delimited, which first is largely unknown to us and can only be discovered through further research. In that sense, the universal law of nature without ifs and buts is an illusion. This can not be accepted by 'science' for the purpose of justifying an ideology or a religion.

Consequently, in this dialogue the observer with his errors and biases is the problem. We talked about space and time in the last chapter.

From a biological point of view, we are *eye-animals*, which means that the visual sense is much more developed with us than the other senses, which is why our knowledge is shaped primarily by the visual sense. The advantage of this sense is its long range, the disadvantage is that it is therefore very easy to fool. The Hindus have invented their own deity for it, Maya the power of deception. To escape this power, we have to deal with the imaging process.

We need to separate the physical reality from the delusion of our senses.

3.1 The Imaging Process

The imaging formation in the eye of the beholder is the same as with a camera. The space is therefore perceived on the retina of the eye as a two-dimensional image with the corresponding perspective distortions. In addition, the perceived image depends on the viewer's viewing location. An image is always location- and direction-dependent. Consequently, you can get completely different images of an object without being able to assign them to the same object, unless you know the way the camera has taken.

To illustrate this, let's take a simple example: we want to record a house by a movie. First we look for a suitable place of observation from where we have a good overview of the house. From there, we will use the camera to walk around the house to register the back. We may also need a flight with the camera over the house. So we will be able to grasp all the views of the house.

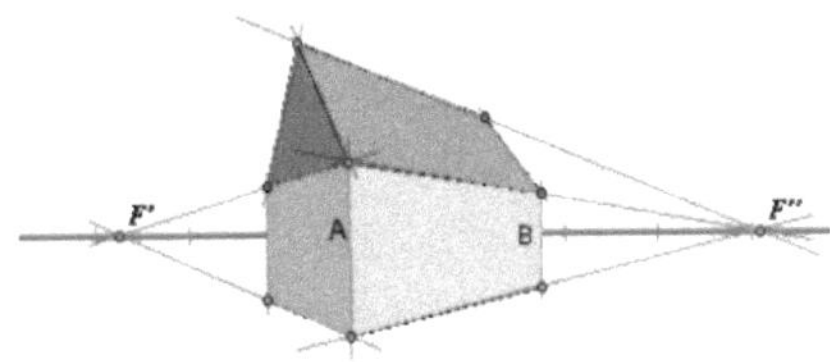

Figure 3.1: Projective transformation - imaging principle of a camera

Although many single images will be similar, we will get at least 9 distinctly different images. All these images are called mathematical projective transformations and the imaging process on the movie is called, how can it be otherwise, perspective illustration.

Our camera delivers pictures like Figure 3.1. You obtain such an image by central projection. In the central projection, the projection rays intersect at one point, the projection center Z. The image plane of figure 3.1 is equal to the *xy*-plane at z = 0 and the projection center by a lens lies on the negative z-axis at point Z = (0,0, - *a*). It goes without saying that the projection center must not change during the image formation. Given a point *P* = (*x, y, z)* on a house edge. We are looking for the coordinates of the projected pixel *P'* = (*x', y ',* 0) on the image plane. Then it follows from the set of rays:

$$x' = \frac{x}{1+z/a} \quad y' = \frac{y}{1+z/a} \quad z' = 0 \qquad (3.01)$$

Such images have been familiar to us since childhood. We consider the edges A and B and recognize that the edge A is larger than the edge B in the picture. After a series of pictures on the film, we reached a position opposite the house edge B. Now the house edge B is shown larger than the house edge A on the image. If we still wondered about it as a child, today we do not think about it anymore. Since this contradiction was caused by the change of the point of observation, we know that our observation deceives us of the true facts. The edges A and B are the same length in nature. Only the depth lost by the transformation into the image results in the need for 'vanishing points' F' and F" on the horizon for symbolizing parallel lines in reality to indicate the depth; because we know that a more distant object always appears smaller than one less distant object. Our eye is as constructed as the camera. The distortion is literally in the eye of the beholder and not in reality. Because a three-dimensional space is transformed into a plane, distortions occur on the image. Mathematically, this is seen by the fact that the vectors $\boldsymbol{x}'$ and $\boldsymbol{y}'$ have each become functions of the coordinate z. Thus, the spatial relationship (2.01) with nonzero values of $\boldsymbol{x}'$, $\boldsymbol{y}'$ and $\boldsymbol{z}'$ applies here no longer. We get with (3.01):

$$\boldsymbol{x}' \cdot \boldsymbol{y}' = 0 \tag{3.02}$$

All that remains in equation (3.02) is the scalar product $\boldsymbol{x}' \cdot \boldsymbol{y}' = 0$ with non-zero values, since here the cosine between the two vectors $\boldsymbol{x}'$ and $\boldsymbol{y}'$ *i*s still zero. The coordinate z‘ disappears in the image.

3.1.1 How to Describe the Change of the Observing Place with Mathematics?

A change of location of an observer or object is called in mathematics an Euclidean transformation. But mathematically, it is equivalent whether you move or turn the observation site or the observed object. Energetically that is not the same. A physical transformation always involves the consideration of the masses of the observer and the observed object. But if I move the object or go to another view point, I have physically done a work, but this work is disregarded in the mathematical transformation. If I move an object in the visual space, which is also called the Euclidean space R^3, it has no influence on its shape. Only the coordinate system, the frame, is shifted or rotated relative to the observed object, that means I have moved. This changes the location coordinates in the room. This is different with projective mappings or projective transformations. There, the shape changes too so that only the linearity is retained. The angles and lengths change. See Figure 3.1.

3.1.2 The Mystery of Lorentz Transformation

Now let us examine a transformation that has become known as the Lorentz transformation and that Einstein has encouraged to his theory of relativity. This transformation looks like this:

$$t' = \frac{t-(v/c^2)\cdot x}{\sqrt{1-v^2/c^2}} \qquad x' = \frac{x-v\cdot t}{\sqrt{1-v^2/c^2}} \qquad (3.03)$$
$$y' = y \qquad z' = z$$

If we compare this transformation with our equations (3.01), the similarity immediately stands out. But now we have one more dependent variable, the time *t'*. Equation (3.01) deals with the functions *x'* = *f (x, z)* or *y'* = *f(y, z)*. The equations in (3.03) are the two functions *t'* = *f (x, t)* and *x'* = *f (x, t)*, and *c* remain constant like *a* in (3.01). Functions with two dependent variables describe surfaces that are mapped to a straight line. This fact differentiates these equations from Euclidean transformations in which there are no functional links between the coordinates that lead to projective distortions that retain only the co-linearity. The first mapping function causes an observed shortening of the time for the moving object in the direction of movement relative to the stationary observer, and the second mapping function causes an observed shortening of objects in the direction of motion as a result of the functional linking of x and t, relating to the resting observer. The meaning of the projection center a from the equations (3.01) assumes in the equations (3.03) the square of the speed of light, which is why Einstein has demanded the constancy of the speed of light in a vacuum. These distortions are exactly the same, which an observer observed at the receding object as a result of perspective. The only difference is that the perspective optical illusion is recognized by our two eyes as a spatial depth. This is the subject of stereometry. On the other hand, we have no practical experience with the effect of the equations (3.03). We would also have to choose two sites to get objective statements. We see that the Lorentz transformation belongs to the projective transformations.

It is highly unlikely that a real observer will ever experience this 'Lorentz perspective', since under normal conditions cosmic massive objects can barely exceed one per thousand of the speed of light, unless they are ionized and accelerated in electromagnetic fields, dissolving the form of the Observers. If you were to use the speed of sound instead of the speed of light, the distortions in (3.03) would increase even more. At least now you would have to be suspicious, because the speed of sound is reached by aircraft, and no distortion of pilots were reported. We state:

Observations of objects are subject to optical illusions, when the observed objects are far away.

3.2 What happens to the Observation?

Observation is a relation between observer and observed object. When you go around the house with the camera and take pictures, you can take a lot of pictures of the house. As a healthy person you will not save all the individual images in the long-term memory. That would overwhelm our brains. Therefore, we compare the observations and examine them for common features and summarize them. In nature, biologists search for life forms, compare these forms and classify them into feature classes, into a system that they provide with terms. Chemists are interested in substance classes. Often, smells and colors also play a role for them. Physicists are interested in what they can quantify and classify by units of measure. In any case, the classification of observed objects serves the purpose of concept formation. The term 'class' is synonymous with the mathematical term *set*. Just as we can group sets into a new set, concepts can be grouped into new, more abstract terms.

For every object or life situation we have a lot of pictures and other sensory impressions. We learn concepts for that. That happens when learning the native language. It seems so obvious to us, later that we do not think about it anymore. We are accustomed, in the common sense of visible things with form, to call as objects. The objects of physics, on the other hand, are quantitatively determinable properties of matter. What properties these are, is given by the physical system of measurement, to which we will return in section 3.7 Modeling, Experimental Design and Measurement. These properties are not detectable only with the sense of sight, which is why the form for the physics does no matter, especially as it knows no quantitative measurement method for determining the shape.

A scientific term always represents a set of objects or sensory experiences with unique characteristics. A basic physical term is also represented by its unit of measure. When I speak of basic concepts, there are also other concepts that are composed of basic concepts. In fact, at a higher level of abstraction you can form higher terms from the basic terms of a scientific discipline. For example, matter is such an abstract term, consisting of the terms energy, mass and velocity or energy, effect and frequency. A scientific term thus always consists of a series of clearly observed characteristics and sub-terms. Immanuel Kant, the great philosopher of the Enlightenment, spoke of *phenomena* in contrast to *noumena*, meaning a term like God, for which there is no sensory experience. A number of terms have been included in physics that have no sensory experience, such as black holes as mathematical singularities, dark energy, dark matter, or antimatter and quarks, to name but a few. These terms can then be equipped with any features. Usually in such terms the features change according to the needs, as well as the meaning of the joker in the card game Rummy. A typical example is the term of the black hole. It was originally a mathematical singularity in the equations of General Relativity, a gravitational anomaly that should swallow all matter, including light. Nowadays the astrophysics want to have found a black hole in the brightest part of a galaxy, spitting out plasma jets, because they believe that a mathematical idea must correspond to a real physical object. It just had to be discovered, like Godel meant.

While the scientific approach of conceptualization always starts from the classification of phenomena, nowadays we often encounter the pseudo-scientific method that invents a term and searches for phenomena in the real world. This is exactly what Immanuel Kant criticized in his major work *The Critique of Pure Reason* [3.02] as early as 1781, for which he was placed by the Catholic Church on the Index of Forbidden Books. Kant showed that we only know about things insofar as they reflect on the sensory perception in our minds. The 'things in themselves' are only approximately recognizable by their appearance. Today we talk about models or concepts about real things. Let's take a familiar example. Everyone has an idea of a house. This idea may be completely different for any person, as there are very different houses. In the future, houses will be designed by architects and built by builders, none of whom can presently have an idea. Nonetheless, there is a mental model of the term '*house*' that will bee accepted as such by all people, based on their essential needs for a home. We will talk about modeling in section 3.7 Modeling, Experimental Design and Measurement

3.3 Knowledge and Faith

As a child I learned to follow the instructions of my parents and teachers because I had learned that they are teaching me to my benefit. I believed in her authority. That changed with growing up. I noticed that I could not rely on the advice of my parents anymore. Belief in their authority vanished. I had made my own experiences.

Faith builds on instruction. The knowledge based on experience. According to Kant, however, knowledge has its clear restrictions where sensory perception ceases. I would like to extend this sensory perception by measurement techniques. You can not know anything about things that aren't detectable sensually or by measurement techniques, even if there are always people who presume to be able to provide information about such things. Such things are empty, Kant says. They serve as projection surfaces of our fantasies, fears or wishes and so we fill them with our ideas. Examples include gods, Dante's Hell, the Big Bang, black holes, fire-breathing dragons, and other curios. Knowledge as sensory experience is defined as a justified opinion in the social context about an object, which includes facts and rules and theories. But there is no guarantee that this opinion will remain invariably true. In science, there have been countless cases where an opinion had to be replaced with a new one because it could be falsified. Truth is not absolute, as we learned in section 2.3.1. What is Truth?, only an assessment. Facts, on the other hand, are always true because they do not spring from inductive logic but come from immediate perception. If perceptions contradict opinions, it is advisable to correct the opinion rather than try to deny the facts because of authority.. A scientific opinion is therefore still no knowledge. Knowledge can only emerge in social development. But that requires a doubt. The doubt, however, is in contradiction to belief, which is basing on the authority of the mediator. Thus, even my subsequent discussions and doubts are not yet knowledge, as long as they have not found social recognition, even if they build on knowledge, they remain in the hypothesis stage. They merely offer an alternative to current hypotheses about observed phenomena, for which there are supposedly no alternatives.

If the truth content of opinions is unknown or controversial, it is called an individual belief. This belief, as in the example of my parents, does not have to be religiously motivated, but it builds on authority. Scientists and for many people also priests are accepted authorities. Religious beliefs always include larger groups of people. In Christian circles, you can currently find the opinion that physics has ensured that the confidence in the mind (science), which has grown since the Renaissance, has dropped again. Not only Christians would struggle with the impertinence of their faith. Even atheists would have a problem to set the rationality absolutely. That does not work anymore. -- Is there then a relative mathematics? Our computer-implemented logic assesses with only two values. There are also multivalued logic's. But these are not integrated into our digital world. In the border areas of physics they might be useful. The mathematics familiar to us also asses, but with ten values. Most people are unaware that logic is the algebraic basis of all mathematics. If a physical model is illogical, it contains a mathematical error. Nobody of the model builders can then claim that we only wouldn't understand this model. Then it has to be worked on until it is understandable. From incomprehensible things people turn away and they are looking for simpler explanations. Science suffers only a loss of authority because it is not up to date. The preference for simpler solutions goes back to Wilhelm von Ockham in 1341. In his best-known formulation, the Ockham principle comes from the philosopher Johannes Clauberg. He wrote in 1654: "*Entia non sunt multiplicanda sine necessitate*" (English: "entities must not be multiplied without necessity.") [3.03]

The fact that observational results can be different is due, on the one hand, to an often overlooked aspect, namely the place of observation, the observation method, the previous education and, on the other hand, the mental attitude, the consciousness.

> *»Consciousness therefore from the outset, is been a social product, and remains so as long as people do exist.«* *- from Die deutsche Ideologie*

These aspects influence the observation result and this influence must be considered in the evaluation. It is especially good to study the influence of a mental attitude on the image of the ancient Egyptians. Probably no folk of antiquity has better documented its way of life and belief on durable material. The ancient Egyptians did not yet know the laws of perspective, but they had a clear idea of space and surface. This is proven by their depictions. While the ancient Egyptian sculpture clearly captures the spatiality of a figure, the image on a wall surface is a bit strange for our present viewing habits. Everything, both human and animal in motion, is represented in profile, while strangely, only the eyes face the viewer head-on. This contradiction is explained by the function of the picture. This is a narrative representation of the depicted persons. While the movement can only be shown in the area, the eyes turn to the viewer because they want to tell him something. We see that the images of the ancient Egyptians are of great symbolic power, culminating in the hieroglyphics, which is why this interpretation is quite likely. The observer translates his observation of the real world into a model - a series of scenes, words and symbols - that can only reproduce a part of reality. That's the part he considers worthy of observation. This ability is innate to humans and therefore does not seem worth considering. But it is fundamental for the data acquisition of an observation.

Observation is not just an image recording but also filtering and classification.

It is filtered and classified only by the observer, what he understands and what he considers essential. Because the observer only consciously observes what he understands, observation is not only a function of perception but also of the observer's knowledge. Since knowledge is not based solely on the experience of an observer, but includes social experience, the knowledge is much more extensive and the observation can be classified in this wealth of experience. But his faith also flows in. Only the statements about these experiences and their socially consistent evaluation about logical reasoning steps leads to knowledge. If this rating is socially contradictory, it is called belief. In this respect, there are reasonably reliable criteria by which you can distinguish knowledge from belief. For this purpose, the observer uses the logic of his mind. However, there are basic assumptions in every discipline of knowledge that can't be reduced to simpler statements and therefore are not justifiable in the specific discipline. Therefore, the aim of any knowledge discipline is to reduce the basic assumptions to a minimum.

These basic assumptions resulted from a higher discipline from the past. That was the philosophy. Philosophy was considered universal science over the centuries and was more or less linked to religion, which is why we should not look only at the philosophy of one religion. Religions are mirror images of the societies that produced them, and so they bear ancient experiences of these folks. But we should not necessarily take their statements literally, as they always served the power interests of their representatives too.

With the progressive independence of the individual sciences in modern times, it has become increasingly problematic wanting to operate a universal science. Since Heisenberg, theoretical physics even claims its own philosophy with its own logic, the consequences of which are yet to be considered.

The more we are coming to the limits of science, the more we have to work with unproven assumptions. But more often, these assumptions should be disputed and questioned. This is uncomfortable, which is why we most rely on authority. However, these assumptions should by no means violate the four fundamental *principles of physics* derived from philosophy.

The first basic principle is the preservation of matter

This means that neither mass in the universe can be created from nothing or destroyed, but that the moving mass corresponds to an energy and vice versa. At the same time, however, there is also a speed limit for the movement of masses, namely the propagation velocity of the light, which does not express this equivalence. While we do not yet understand all of these processes, we should be guided by this ancient fundamental principle.

How important the philosophical attitude is for a natural science is shown in an article by Paul Marmet titled *Absurdities in Modern Physics* [3.04], dealing, among other things, with the law of causality and the interpretation of quantum mechanics. It was overlooked that the quantum-mechanical approach is a statistical one aiming at data compression and not considering the functional relationship between cause and effect, but the causal law is objective and neither dependent on the viewer nor nullified in the microcosm, so as Heisenberg claims in his essay '*Quantum mechanics and Kant's philosophy*' [3.05]. If the causal law were abolished in the microcosm, you would have to find a limit from which the causal law is abolished, but it is the statistical approach that eliminates causality. This view is expressed in the thought experiment that has gone down in the history of physics as *Schrödinger's cat* [3.06]. The background is the radioactive decay of atoms, the cause of which is unknown and describing in quantum mechanics with a wave function that does not presuppose a cause. Quantum mechanics is a theory in which a description of the micro-world takes place on the basis of reduced information. Here the statistical description replaces the causal description. The ignorance of an effect does not mean that there is none. The 'thing in itself' is pretty unknown.

Western culture thinks causality as a process controlled by others, with an external cause. The Far Eastern culture, on the other hand, thinks of causality as an autonomous process, where the cause lies within the process and the effect creates a feedback. This can be seen on religious beliefs: "*The Lord shall guide me*," or "*so God help*". In Far Eastern culture, the wheel of life is at the center of religious worship. Cause of future fate is acting in the present. The 'karma' determines destiny. Good deeds improve karma, bad deeds reduce karma and chances in future life. Every human being is responsible for his karma.

The second basic principle is causality: It says there is no effect without cause. Causality is usually seen linearly. In autonomous systems it is cyclical.

Many physical systems are autonomous. An example is the spread of light. Science and in the narrower sense the physics, is based on observations and experiments. While most of the observation is done without any impact on the environment, the experiment is the immediate planned influence on the environment to observe an expected event in the form of a movement or change in the state of an object. Each experiment is based on a causal link between cause and effect.

The experimenter, while preparing the experiment, strives to dissect the causes for the effect he intends to achieve. In the experiment itself, independently varying characteristics caused by the experimenter are measured and a hypothesis is made on the effect on a variable suspected to be dependent, and this hypothesis is compared with the measurements of the latter variable. When repeating one and the same measurement, it is observed that a different value is read each time. We speak of random deviations from the mean value.

The mean random deviation from the mean value over a series of measurements is an important measure of the reliability of the measurement method. Without the law of causality, there would be no justification for establishing a hypothesis about a functional connection. Therefore, the causal law on linking cause and effect is so fundamental to the scientific method that a violation must be rejected as unscientific. Mathematically, the causal law is a function between an independent cause and a dependent effect.

Effect = f (cause) or y = f (x)

We consider the movement as a time-dependent quantity of the way, which is why the effect always depends on the cause and not the reverse. The effect unfolds from the temporally preceding cause, whereby some processes can proceed at the speed of light.

Physics as a natural science is not concerned with the mind of the observer, for this psychology and neurology are responsible, but with the movements of matter and its causes. Mass and energy are the most important forms of matter. Derived from the causal law, the basic assumption of physics is that matter can neither be created nor destroyed. It can only be transferred to the various forms of existence. Any change of a form of existence, however, provokes a backlash.

A third basic principle is the unity of opposites.

The world is polarized. It is based on a dynamic balance. The world is in harmony then. In Indian philosophy '*Vishnu*' has as an opponent, named '*Shiva*'. This principle can be demonstrated in many cultures, for example, to trace back to the Chinese 'Dào'. Dào with its *Yin* and *Yang* means as much as the way, but also method and principle. Dào describes in Daoist Chinese philosophy an eternal principle of action. The basic idea of this philosophy is that all existence arises from the lawful transformation of the basic forces 'yin' and 'yang'. These dynamic forces sometimes compensate each other more, sometimes less, and the goal of human action is to bring them into harmony, into balance. These ideas extend into the 10th century B.C and are already contained in the oldest Chinese philosophical writings.

»All things under heaven arise in being. Being arises from nonbeing.« - Lao-tse

But nonbeing can only be recognized in being. Or how should you recognize a hole if it had no edge? How would you like to identify a positive charge if there was no negative and there was no force field in between? How could you distinguish a high potential from a low if there would be no energy flow between the two? Because of these opposites, everything in the world is in motion. However, this principle already contains a number of misunderstandings. You could conclude from this the symmetry of the world. Symmetry is a static balance that contradicts life. Although there are positive and negative charges, these are distributed over masses of different sizes, which is why we can not speak of symmetry but only of similarity. Another misconstrued term is the materialistic dialectic used for the unity of opposites. The hereon constructing social theory has failed, in allowing of free forces. Similarly, in physics it is believed that paradoxes are logical contradictions expressing a dialectic. The term comes from the Greek and means art of conversation and it should be used only in this context.

As mentioned earlier, Einstein introduced symmetry as another belief. But that was not helpful and that destructs scientific thinking.

We replace the symmetry by self-similarity as a fourth basic principle of nature.

Self-similarity is a principle of repetition of structures on different scales with insignificant modifications. Later, in the context of structure formation, we will discuss the principle of self-similarity.

There are two unfruitful beliefs in physics

1. The belief that two different ways of looking at physics, a deterministic one and a statistical one, can unite and thus achieve a unified physics. In fact, today's modern physics is breaking away from the classical, without taking a step closer to the dream of merging the two ways of looking at things.

2. The belief that you can get more insight from the mathematical model than you put into it. That would be a mental perpetual motion. With the expert systems of the artificial intelligence they realized very fast that one does not succeed. A computer remains a speed dork. It does it all wrong very quickly, which was not given correctly to to the machine. Computers also have to learn from the environment and that is for theories the same. To acquire knowledge, the systems have to be open.

3.4. Watching Aspects

What is a *watching aspect*? There are many synonyms for method of approach and aspect. I choose the term watching aspect because it has not been used yet and I can set its meaning here. Under a watching aspect, I want to understand how and viewpoint to observe a phenomenon. It is practically a prejudice that I decide about order or chaos in my field of observation. If I am interested in the causal relationships, this is a deterministic aspect of observation. My observation objective is the description of my observation in the form of functions and equations. There is another aspect of observation that involves non-deterministic consideration. This approach assumes that there to one cause can be multiple after effects. Often a statistical approach is the first choice to get an overview of which questions to ask first. A statistical approach is a kind of data compression that takes no account of causal relationships. The two aspects of the analysis are in an ambivalent interrelated relationship and, in addition, there are some aspects of data compression in non-deterministic consideration.

3.4.1 Causality and Coincidence

If we observe a phenomenon, the question immediately arises as to its cause. However, if the question can not be answered, God usually comes into play for believers. For all ignorance gods had to serve practically as Joker. The joker for the unbelievers is coincidence. While God leads into the realm of theology and the causal law is abolished there by miracles, coincidence leads to statistics, where the fathers of quantum mechanics also thought that the causal law no longer applied there, as we have already discussed above. However, statistical methods only serve to limit the number of influencing variables in the mathematical modeling of natural phenomena, influencing factors that can not be grasped.

That causality and coincidence are not contradictory will be demonstrated below. We humans think in causal chains. We line up causal strands in time, just as we write a text. But what happens when causal chains influence each other?

To investigate this, we need an example: Suppose a board of square cells! On the game board, we randomly distribute only black tiles into arbitrary cells as initial state. Now we need a few rules. Rules of the game are, you will guessed it, causalities. We play the *game of life* as first played by J. H. Conway [3.07]. Because it being a game, we are completely free in inventing game rules. Consequently, they have nothing to do with real life. Nevertheless, you will see that you can learn a lot about causality from this game. We can play the game completely alone on a game matrix with monochrome tokens, before we come up with further rules for dealing with tokens of different color.

We need rules for adding and removing tokens or tiles. The addition of tiles is what we call the birth of an individual, and we call the removal an individual's death. Each tile has an environment. These are the eight neighboring cells around each cell may be occupied by a tile or not, border cells excluded. The existence conditions of each tile depend causally on its environment according to established rules of the game.

In a game generation, the rules of the game are applied to all cells in the field from top left to bottom right in turn. The game can continue for so many generations until it reaches a state that makes it impossible to continue playing, or achieves a periodic stability that consists in two states alternating over and over again.

But now to the individual rules of the game for the existence of a playing piece:

1. If the environment of an empty cell has three neighbors, then an individual is born, i.e. a tile can be placed on the appropriate cell.
2. If the environment of an occupied cell has only one neighboring tile, then the owner of the occupied cell dies of loneliness. This tile must be removed from its cell.
3. If the environment of an occupied cell contains two or three neighboring tiles, then the owner of the occupied cell remains alive. This tile remains on the field until the next generation.
4. If the occupied cell environment contains more than three neighboring tiles, then the owner of the occupied cell dies of overpopulation. The tile must be removed from the cell.

Now we have to set up a rule for the fringe cells, since their environment has only 5 or in the four corners only 3 cells. There are basically three options: the missing cells are either considered empty or occupied or you think the playing field is self-contained without edge and you look at the fringe cells of the opposite side. This is a bit complicated in a corner cell, because for each corner cell 2 cells of the opposite corner must be considered and a diagonal cell.

Playing the game in the manner described is undoubtedly very tedious over time. It is easier to hand over this task to a computer, to expand the game matrix and to have the order of the tiles displayed after each generation. It will be astonishing how varied the patterns will be, depending on the amount and starting order of the tiles.

There are orders that have died out after a few generations. Other arrangements move from generation to generation across the field, others grow before disappearing from the playing surface. Particularly fascinating are the moving and oscillating objects, because they remind of the propagation of electromagnetic oscillations.

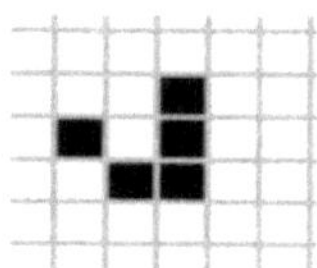

Figure 3.2: Play pattern glider (animated) Reference: WIKIPEDIA

Although all game arrangements are based on 4 simple rules, on causality, it is not possible to predict how a given initial arrangement will behave. You have to play it through to find out. In the example from Figure 3.2 it works cyclically. However, if you have another initial condition, then a completely different behavior can come out. We are confronted with such systems in physics more often.

What happened here? The four rules are linked to the game-field and they act no longer independent, but coupled depending on the behavior of their neighborhood and this neighborhood has two different states, occupied or free, which in turn depend on their neighborhood. If you leave the edge of the game-field, the environment eventually even depends on the neighborhoods of the previous generations and it comes to feedback until the dependencies become so complex that they are no longer manageable for the human mind. Mankind is not trained by evolution to master complex situations of interdependence based on causal links. We help our-self by declaring such phenomena as random.

Mathematics not only has linear dependencies but also non-linear ones. Nature is rarely linear. For example, if one determines the roots of a nonlinear polynomial, then you obtain as many solutions as the degree is of the polynomial. Which solution nature chooses is random from the standpoint of mathematics, since the selection criteria have nothing to do with mathematics, but with the environment in which the nonlinear process takes place. Therefore, for a cause, nature often offers different effects that may unfold differently in different environments. The simplest example of this is the fall of a cube.

In complex systems, a priori, we can not decide whether an event is randomly or causally autonomous.

An example of causally autonomously, is crystallization. From a liquid with seemingly chaotically moving atoms and molecules, a regularly built body crystallizes. There is a *self-organization* out of chaos here, which sheds a whole new light on our understanding of nature at phase boundaries [3.08]. The term self-organization is the term for a class of natural phenomena, which obviously have an inner autonomous set of rules, without it immediately opening out to the observer. For engineers, such control systems are typical building blocks of automation. Many systems, formerly regarded as chaotic systems, are today recognized as well-ordered. The necessary modeling measures are still the subject of research. It forms here another aspect of watching out. Some people even talk about a necessary paradigm shift. But that would mean losing sight of the other watching aspects.

3.4.2 Determinism versus Statistics

Strict determinism is the illustration of the epistemological and scientific theoretical view according to which it would be possible in the sense of the idea of a closed mathematical system of world equations under the knowledge of all laws of nature and of all initial conditions such as the position and velocity of all physical particles present in the cosmos to calculate and determine each past and future state. According to this statement, it would be theoretically possible to set up a *theory of everything* (TOE). This determinism is referred to as 'Laplace's demon' according to a quotation from Piere Simon Laplace of 1814.

»We ought then to regard the present state of the universe as the effect of its anterior state and as the cause of the one which is to follow. Given for one

> *instant an intelligence which could comprehend all the forces by which nature is animated and the respective situation of the beings who compose it -- an intelligence sufficiently vast to submit these data to analysis -- it would embrace in the same formula the movements of the greatest bodies of the universe and those of the lightest atom; for it, nothing would be uncertain and the future, as the past, would be present to its eyes.«*

Although Immanuel Kant thirty years before Laplace already pointed to the impossibility of complete knowledge, this demon still haunts through the minds of theoretical physicists. The idea of a Theory of Everything still haunts through the parlors of scholars. Theoretical physicists find it difficult to get used to statistics. »*God does not play dice!*« was an apt phrase from Einstein to characterize this mentality. But our minds are overwhelmed by the complexity of the world, so we have to do a data reduction and that's what statistics are for. One reason for the success of Laplace's demon in the 19th century was the generalized description of dynamic systems by the Hamiltonian function. This is nothing other than the sum of the potential and the kinetic energy of a closed system, which are no longer described by places and velocities of the components of the dynamic system, but by so-called canonical variables: the coordinates and impulses, which are denoted by p and q. This function is represented as Hamilton's function $H(p,q)$ and should contain all the empirical knowledge about the system. But it does not, it contains no interactions as we have modeled them in the game of life. You can think of the Hamilton function as a film whose individual images depict the p and q at a given moment in each case, only that physicists can not imagine state descriptions, this requires an engineering education. The above theoretical approach proved to be one the most successful in the history of physics. These canonical equations are symmetric in terms of time and velocity. According to this model, we can determine the future as well as the past, because the elements do not interact with each other. In the previous section we have just seen that this approach does not work in practice because it does not take into account the interactions of moving particles and interactions are abundant because of the force fields of the particles, otherwise quantum mechanics would not have been invented. Theoretical physicists find symmetry beautiful. (There are other beauty notions – e.g. beauty is the promise of function.) We had already discussed the fateful overvaluation of symmetry in the previous chapter. Because of the Hamiltonian function, symmetry plays such a big role in theoretical physics. Now the temporal change of the impulses depends on the derivative of the Hamilton function according to the coordinates. For free particles, the derivative disappears and the pulses become constants. In our imagination we get a single picture from the movie. However, this is a very dramatic requirement for the mathematical approach, the importance of which has often been underestimated and limits the applicability of this mathematical approach to linear systems. This has become apparent for the first time in the three-body problem in celestial mechanics. The three-body problem of Henri Poincaré is to find a solution to the trajectory of three celestial bodies under the influence of their mutual attraction. To obtain quantitative results, it had to be solved numerically in the general case so far (by approximations). The statistics provide a completely different approach.

> Imagine that you are sitting in the SCOOM at one end of Alexanderplatz in Berlin and you realize that you need a pair of new shoes that you could possibly get from Reno on the opposite side. Nothing easier than that. You cross the Alex on a straight line. But just this day is market. A lot of stalls are built up and crowds of people walk around in between. In order to reach Reno, you must constantly dodge people or walk around stalls on your way, keeping a certain distance from the obstacles. Since you have a modern mobile phone with GPS function in your pocket, your way is recorded and you can watch it later. This path is not a straight line but a curve more or less reminiscent of a wavy line. If you want to reach the starting point again, your path will certainly not be the same, but also of undulating shape. All people will be able to make the same observation as they go this way, but there will not be two paths that are identical. If you want to describe this way in general, you can confidently take a simple sine wave without making a big mistake. This overlay is written as $|\psi\rangle$ with ψ as the **wave function**. The wave function itself has no physical significance. It is merely a symbol of the indefiniteness of the whereabouts of any passers-by.

In the microcosm of atoms, one constantly encounters such a condition as "market on Alexanderplatz" due to the thermal motion of the atoms. Due to the force field around the particle, only certain distances of the passage are possible here. Therefore, in quantum mechanics, the dynamic variables of classical mechanics are replaced by linear operators that work on a general wave function. It is declared that the same identity relations between the operators as in classical mechanics. Now you can ask, why all this effort? Waves have discrete resonances and these are useful for the description of discrete energy levels in the atom. The mathematics is very demanding and is not subject matter at high school. The whole thing is shifted to the area of complex numbers. Only by squaring does you get back the real part of the wave function. This is assigned a physical meaning as a residence probability. This works very well in the description of the electron orbits. However, I would reject the quantum mechanics within the atomic nucleus, as it must have a solid structure that acts on its shell to guarantee the tight binding properties of the elements. A quantum mechanical model of the atom nucleus did not really make us any wiser.

A static approach to complex operations is quite legitimate when dealing with large amounts of particles and having to perform data compression to describe system behavior. However, it becomes nonsensical to try to unite the statistical description with the deterministic description, as attempts show to create a theory of quantum gravity. Physicists were trained in solving mathematical equations at least since my studies in the sixties. But they have not learned what their equations actually describe, otherwise they would have noticed that such a project is doomed to failure. One of the dead ends was string theory, which dealt with maximally two-dimensional structures in nine-dimensional spaces. Another meander is the loop quantum theory. It considers space and time on scales of the Planck constant (10^{-20} fm) [3.09]. This is 20 orders of magnitude less than the proton radius and beyond any technical measurement perception.

3.4.3 The Uncertainty Relation

Werner Heisenberg [3.10] discovered that the energy change in a time window is always larger than Planck's constant of activity per wavelength. This discovery is based on the Fourier transform of the waves taken with a spectrograph. Described as the Uncertainty Principle, this discovery is considered to be one of the most fundamental aspects of the Copenhagen interpretation of quantum mechanics. He claimed that the uncertainty principle is not the result of technically recoverable imperfections of a corresponding measuring instrument, but of a fundamental nature. This was sold as a great scientific knowledge and regarded as a natural law.

$$\Delta E \cdot \Delta t \geq \frac{h}{2\pi} \qquad (3.04)$$

Formula (3.04) describes this circumstance, where: *ΔE* is the resolution in energy, *Δt* is the resolution in time, $h = 6.6 \times 10^{-34}$ Js is Planck's constant, which is the energy quantum for the energy emission of an electron per wavelength in the atom. If we substitute $\Delta E\ \Delta t = \frac{1}{2}\ \Delta p \cdot v \cdot \Delta t$ in formula (3.04) we obtain $\Delta p \cdot \Delta x > h/\pi$ or else

$$2m \cdot \Delta v \cdot \Delta x > \frac{h}{2\pi} \qquad (3.05)$$

In other words, this means for quantum mechanics that you can not simultaneously determine the location and the momentum arbitrarily. That this is a law of nature is an unproven claim. Paul Marmet contradicts [3.11].

> *»The limit suggested by Heisenberg can now be interpreted either as the instrumental boundary of a spectroscope or as a mathematical limit because of the loss of information in applying the Fourier transform without the imaginary part. In the case of a phase-sensitive detector, there is no fundamental limit. With a phase-sensitive detector, the limit of resolution depends only on the threshold caused by the noise level, and it is unfortunate that It is unfortunate that Heisenberg could not distinguish between the information loss caused by an instrument and a fundamental physical principle.«*

Every camera maker knows the phenomenon that the sharpness of a shot depends on the ratio of the opening time of the lens to the moving speed of the picked-up object and the focal length of the lens. Blurring is a feature of the observer's technique and is caused by insufficient technical equipment in the observation. It is not a law of nature. We will come back to the consequences of Heisenberg's statement under Chapter 5 in connection with a particle property.

The fact that the mass is contained in (3.05) leads to the conclusion that we can also define quanta of effect with arbitrary other masses. We will deal with a quantum of effect with another mass in section 3.4.5 Planck's Constant and the Domino Day . But first let us ask the question about the general limits of observability.

3.4.4 Limits of Observability

In general observation would begin with our sensory performance, if engineers would not have upgraded our sensory performance through appropriate technical equipment. For all our senses, we have equipment available that allows us to amplify and transform our sensory endeavors. We therefore have to ask, where are the limits of the used devices? Basically, the limit is determined by the signal-to-noise-ratio or the optical resolution when it comes to waves. The signal-to-noise ratio can be defined by the ratio of the amplitude of the useful signal to the standard deviation of the noise signal. Optical resolution is the distance that two structures in the eyepiece of a microscope or telescope must have at least in order to be perceived as separate image structures after optical imaging.

Let's first take the resolution to perceive two immobile objects in a telescope separated from each other. By specifying an angular distance or by specifying the distance of just separable structures, the resolution can be quantified. For example, the resolution of optical instruments is limited by the diffraction of the light. For both the telescope and the microscope, the limit is about half the wavelength of the observed light. In large entrance pupils of optical systems, the resolution is usually not yet limited by diffraction, but by opening errors of the optics. These can be reduced by changing the aperture, so that results in a critical aperture optimal resolution. For earthbound telescopes, air turbulence mostly limits the resolution to about one arc-second 1". This air turbulence is perceived as sparkling stars. The Hubble Space Telescope achieves a resolution of about 0.05" at visible wavelengths due to the elimination of the disturbing atmosphere. Since the seventies of the last century, instead of a long exposed individual picture, the influence of air turbulence can be eliminated by the superimposition of many short exposed images (speckle interferometry), but this is only possible with relatively bright objects.

A spatial resolution provides the stereometry based on the triangle calculation. If you have a base distance and you measure the angle between the base distance and the observed object from each end point of this distance, you can determine its distance. This is how our spatial eye-sight works. If you observe a star over the year, it seems to make a small ellipse in the sky. But this is not the movement of the star, but the cause is your movement as an observer with which you wander around the sun with the earth over the year. This apparent migration of the star of only a fraction of an arc-second in the sky is called parallax, and the first to determine the distance of a star, was in 1838 after a year of observation by Friedrich W. Bessel in Königsberg. On the basis of this observation, the distance measure Parsec has been defined. It is the distance from which the mean orbit radius (= 1 AU, astronomical unit), ie the mean distance between the sun and the earth, appears at an angle of one arc-second, which corresponds to about 3.26 light years or 206,000 astronomical units or $30{,}9\times10^{15}$ meters. We see that we reach the technical measurement limits relatively quickly if we want to measure the distances in the cosmos directly. The method of parallax is the fundamental distance determination in astrophysics, but the accuracy of parallax measurements by telescopes is limited to angles of 0.01", and thus to about 100 parsec distance, which corresponds to 326 thousand light years.

Solar radius	$6{,}93\times10^{8}$m	2,3 light seconds
Radius of Earth's orbit	$14{,}9\times10^{10}$m	8,3 light minutes
Radius of the solar system	8×10^{12}m	7,4 light hours
Radius of the Milky Way	$4{,}7\times10^{20}$m	50.000 light years

Table 3.1: Size comparisons in the cosmic area

All other measurement methods are based on indirect measurements. All of which are based on assumptions whose reliability is not guaranteed. The next galaxy, the Andromeda Nebula, is believed to be 2 to 2.5 million light-years away. The distance indication by means of spectral redshift is such an indirect method. It is based on the correlation of redshift and absolute brightness. It exploits the fact that the absolute brightness of a pulsating giant star called Cepheid is firmly related to its pulsation period. These Cepheids are still clearly visible in neighboring galaxies. Errors of about 10% and more are quite real for intergalactic range determinations, as the spectral redshift, as we shall see, depends on the density of the intergalactic medium and this density is not constant.

When it comes to observe the smallest structures, then the electron microscope is the device of choice. It can still resolve structures of 0.1 nm. In comparison, the radius of the hydrogen atom is still half an order of magnitude smaller, which is why atoms can not be directly observed. Still smaller structures can dissolve the Mössbauer spectrometer working with gamma rays and when coupled with an electron microscope. In fact, getting down to the range of femto-meters is very difficult. Electrons and protons are on this scale. About the measuring methods themselves and their resolution can be found no publicly available information.

To get an idea of the magnitudes that are typical for the application of quantum mechanics, the following table lists the dimensions of some atomic indications of size.

Wavelength of green light	550×10^{-9}m	550 000 000 fm
Bohr radius of a H atom	53×10^{-12}m	128 000 fm
Electron radius	$2{,}81 \times 10^{-15}$m	2,81 fm
Proton radius	$0{,}84 \times 10^{-15}$m	0,84 fm

Tab 3.2: Size comparisons at the atomic level

This information is unimaginable for us, which is why they need to be enlarged into our imagination world. In order to transform these into our world of experience, we have to increase the above objects billions of times. Imagine that a proton would be the size of a ball with a diameter of 1.7 mm. In the H atom, the electron would orbit the proton at a distance of 32 m. The electron itself would then be with a diameter of 5.6mm, as big as a pea. The wavelength of the visible light would then be comparable with the distance of 550 km, a distance from Hamburg to Stuttgart. The actual size of atoms however, is quite variable and depends on the type of measurement too and can grow in the substance classes by a factor of 9.

3.4.5 Planck's Constant and the Domino Day

Planck's constant became one hundred years old in the year 2000 and is so closely related to quantum mechanics that we can not think of any relation to the everyday world, because the traces are blurred. First we have to explain the term quantum effect. When we use a hammer to hit a nail in the wall, we create an effect on the nail and the wall. But the nail will not penetrate deep enough into the wall on the first blow. We'll have to do more hits on the nail. With every hit, an energy quantum is transferred to the nail, which allows the nail to penetrate deeper into the wall. Now, hammer blows are not defined. A percussion drill is a more precise tool. Now, around 1900, Max Planck discovered that the ratio of the energy of a photon and its frequency is always constant, as in a percussion drill whose frequency can be changed. Armin Uhlmann quotes Max Planck in [3.12] as follows:

> *»...so you find that the migration of energy into the radiation can be prevented by assuming that the energy is in the first place forced to stay in certain quanta. That was a purely formal assumption, and I actually thought so not much over it, but only that, under all circumstances .... I had to bring about a positive result.«*

Thus, he received an important equation of energy, which transfers no mass, but only their action on another mass, but the portioning by an action quantum did not fit into the framework of the former classical physics. Maybe none of these people needed to deal with a hammer until today. Otherwise, Einstein and his disciples might have noticed that electrons actions in an atom shell can not been swept together like particles.

$$h = E / \nu \qquad (3.06)$$

The constant h was called in German 'action quantum', because it is physically an action. In 1923 Louis de Broglie generalized Planck's constant of action by claiming that this constant should represent the proportionality between the moment and the quantum wavelength not only of the photon, but of any subatomic particle. It remains questionable whether h is then really a universal natural constant, since, for example, protons with their much larger mass can achieve a correspondingly greater effect. Although the term is used almost exclusively in the context of micro structures, a macroscopic view is quite possible and offers the advantage of clarity, which can not be said of quantum mechanics.

In the following we want to vividly demonstrate the vividness of an action quantum.

Once a year Domino Day was televised by Dutch television producer Endemol between 1998 and 2009. Huge amounts of multicolored dominoes were set up, arranged into splendid pictures just to knock them over with a single impulse. The goal was to achieve a world record in domino repelling with only a single impulse (linear momentum). If everything was set up with much fondness and effort, it could start. We all sat in front of the TV and waited for the start of the spectacle. Only one stone is knocked over at a starting position and this triggers the domino effect, which, when all the stones are at the right distance behind each other, each domino knocks over one or two adjacent successors and at last all the dominoes are bowling down one after the other.

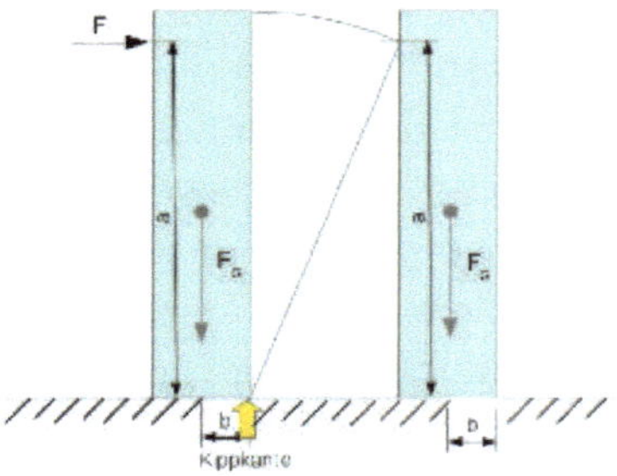

Figure 3.2: The tilting moment

The impulse of the first stone is the triggering impulse that initiates this chain reaction. This impulse is passed from domino to domino while there are still dominoes. By falling over the dominoes, the many-colored images come to light, which were not visible at first. But we are not interested in the pictures, but the physics that makes this spectacle possible. The original Domino stones are 4.8 × 2.4 × 0.48 cm in size and weigh 8 grams. Simply put, we only look at two dominoes a little closer. If we wanted to calculate exactly, we would have to consider the influence of at least one third domino and others depending on the distance of the dominoes, which would unnecessarily complicate matters. Multibody problems can only be approximated, as mentioned in the previous section. Since it only depends on the principle of representing an action quantum and not on its exact numerical value, we leave it with two stones. For a domino to stand, it has a stance moment MS. The tipping moment is responsible for the fall.

The tilting moment ***MK*** arises from the rotational action of the motive force ***F*** with the lever arm *a* about the tilting edge. The stationary moment ***MS*** arises from the rotational action of the weight *FG* which acts in the center of gravity with the lever arm *b*, also about the tilting edge *K*. Thus, a domino falls over, a force F is necessary which tilts the domino by a distance b from the vertical. According to the law of levers, this force can be smaller than the force of gravity acting on the body and attacking the center of gravity. The acting force times the height *a* is the tilting moment ***MK*,** an energy that counteracts the force of gravity times the distance of the projection of the center of gravity on the footprint to the tilting edge. This energy is called stationary moment ***MS.*** If the tilting moment is greater than the stationary moment, the domino will fall over and release the potential energy resulting from the change in altitude of the center of gravity in addition to the overturning moment, and this energy will act on the next domino. So it is possible that a domino can also overturn two more dominoes in parallel. Each domino has potential energy stored in it and at the same time is in a meta-stable equilibrium - the amount of stored energy can be released by the action of a smaller amount of kinetic energy. The release of energy is not continuous, but each stone provides a fixed amount of energy, which results from the gravity of a domino and the change in height of the center of gravity when falling over. We want to approximate the amount of energy that will be released (see Figure 3.3). The center of gravity describes a circular arc around the tilting edge with the radius $r = \sqrt{h^2+b^2}$ when falling.

To tip over a domino, the energy $E = mg(\sqrt{\Delta h^2+b^2} - \Delta h)$ must be expended from. However, since each domino comes to rest on its successor, the change in height of the center of gravity is not the height h_G of the center of gravity of the domino but only $\Delta h = h_G - x$.. To calculate the *x*, somewhat trigonometry is necessary.

$$x = h_G \cdot \sin(\alpha + \beta)$$

The height change Δh in Figure 3.3 then results in:

$$\Delta h = h_G(1-(\sin(\alpha)\cdot\cos(\beta)+\sin(\beta)\cdot\cos(\alpha)))$$

The angles α and β must now be expressed with the parameters h_G, *b* and *d*:

$$\sin(\alpha) = \frac{2b}{d+2b} \quad \cos(\beta) = \frac{h_G}{\sqrt{h_G{}^2+b^2}} \quad \sin(\beta) = \frac{b}{\sqrt{h_G{}^2+b^2}} \quad \cos(\alpha = \sqrt{1-\sin^2(\alpha)})$$

and you get:

$$\Delta h = h_G\left(1-\left(\frac{2b}{d+2b}\cdot\frac{h_G}{\sqrt{h_G{}^2+b^2}}+\frac{b}{\sqrt{h_G{}^2+b^2}}\cdot\sqrt{1-\frac{4b^2}{(d+2b)^2}}\right)\right)$$

$$\Delta h = h_G\left(1-\left(\frac{b}{\sqrt{h_G{}^2+b^2}}\cdot\left(2\frac{h_G}{d+2b}+\sqrt{1-4\left(\frac{b}{(d+2b)}\right)^2}\right)\right)\right)$$

The kinetic energy of a tilting domino is $E = m \cdot g \cdot \Delta h$. Using the above equation for Δh, the energy values 0.0117 J, 0.0139 J and 0.0148 J are obtained for three distances d: ¼ · l, ½ · l and ¾ · l.

As the kinetic energy of a falling domino is enough to overturn several other successors, the chain of events need not remain linear, but can split into any number of chains and thus lead to an increase in events. The chain of events is practically limited only by the number of dominoes placed. In the case of splitting of the event chain, the energy will grow exactly in quantum leaps and be transported further in the new chains.

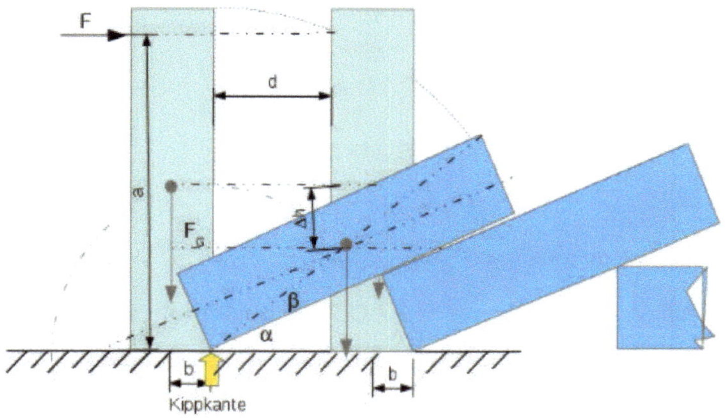

Figure 3.3: To calculate Δh

Now let's ask: how many dominoes per second tip over in a row?

For a domino to overturn a second domino, its top right edge must hit the second stone. To do this, it must cover the height difference Δh_2 = ½ $g \cdot t^2$ with $t^2 = 2\Delta h_2 / g \cdot \Delta h_2$ is the difference of the upper right edge of the domino minus a (Fig. 3.3): expressed in terms of the domino pieces:

$$\Delta h_2 = 2l - \sqrt{4l - d^2} \quad \text{and} \quad t = \sqrt{2\Delta h_2 / g} \qquad (3.07)$$

where *l* is the length of the domino and *d* is the distance between the dominoes. Then we can calculate the number of dominoes knocked over per second. From the above formula we get for

d = ¼ l → t = 0,012s

d = ½ l → t = 0,025s

$$d = \tfrac{3}{4}\, l \rightarrow t = 0{,}038s.$$

We multiply these values by the energy calculated above and obtain for the three values:

$$QD_1 = 0{,}14\times10^{-3}\ Js$$

$$QD2 = 0{,}34\times10\text{-}3\ Js$$

$$QD3 = 0{,}56\times10\text{-}3\ Js$$

as quanta of effect for the domino effect.

For the 'tilting frequencies' $v_i = E\,/\,QD_i$ we obtain with the above calculated value for *E* with 80,7; 40.1; and 26.5 stones per second depending on how far apart the stones are placed. We see here that, when looking at many objects interacting with each other, the energy can be more easily described by a quantum of effect than by incorporating multiple bodies into the calculation of the mechanical determinants. This also gives us a reduction of the free variables as by statistics. We also see that here the momentum transfer depends on the distance of the dominoes. In addition, we find that the impulse was passed on in the force field of gravitation and that this force field exerts a decisive influence on the transmission of the impulse. The quantum of effect proves to be a special way of looking at or describing complex physical processes and that quanta of effect are not physical constants of nature, but depend on the explicit conditions.

It was Albert Einstein, who in 1905, stimulated by Max Planck's law of radiation, in which energy quanta first appeared, came up with Newton's idea that light consists of particles whose energy would be $E = h\cdot v$ with a momentum $\boldsymbol{p} = \boldsymbol{k}\cdot h\,/\,2\pi$ where k is the wave vector. Its direction is the direction of movement of the light wave and its magnitude is $|\boldsymbol{k}| = 2\pi\,/\,\lambda$, where λ is the wavelength of the light. However, as we learned from Domino Day, only the impulse has been carried forward without the particles actually moving away from their original location. Therefore, it does not make sense to consider quanta of light as a particle, just as little it would be considered a hammer blow as a particle. Just as the transmission of hammer blows requires a medium, a medium is required for the transmission of light, a force field whose optical density affects the speed of light. But light spreads in all directions, which is why the wave vector is superfluous.

Whether Planck's constant is really a universal natural constant can be doubted on the basis of these considerations. Pierre-Marie Robitaille and Stephen Crothers has written an interesting article: *The Theory of Heat Radiation Revisited: A Commentary on the Validity of Kirchhoff's Law of Thermal Emission and Max Planck's Claim of Universality,* in which they stated that Max Planck did not use a black spotlight. [3.13]

As we have seen here, no particles have been transported but only the momentum in an existing force field. Another familiar system with a quantum of effect is the AC grid. Unlike DC, where the electrons must be transported over long distances, resulting in significant conduction losses, the electrons oscillate here at 50 Hz. We measure the energy in kWh, no one would think of counting the number of electrons. By division the input energy of a percussion drill by the beat frequency, we also obtain a quantum of effect that is independent of that of the light quanta. Equation (3.06) has a much more general meaning than Planck believed, and the context does not show a constant, since neither the energy nor the frequency are constants of nature, and hence the quotient does not have to be constant. This will happen to us even more often in the following, that the equations do not say, what interpreted into them and thus erroneous conclusions are the result, which is why we do not want to talk about the Planck constant any more, but about the action quantum.

3.5 From Nothing to Infinity

At the beginning of the 20th century there was a dispute between Albert Einstein and Dayton Miller about the existence of an ether as a medium of light transmission, which ended in favor of Einstein. Einstein then tried in vain to develop a field theory to fill in the Nothingness of 'space'. Today, we no longer see the interstellar and intergalactic cosmos as an empty space. Although the cosmic high vacuum contains only a few atoms per cubic centimeter, but that is much more than emptiness, and if the atoms are disassociated in a plasma, it is by no means negligible. It is an optical medium with physical properties, a mass filled volume consisting of gas, electrons and ions of variable density and concentration of its constituents.

Another question is the temporal and spatial limitations of the cosmos. We can not see a temporal beginning or a spatial limit for the volume of the cosmic plasma. This logical dilemma of limited infinity brought Albert Einstein to the idea, to think of the world as a hyperspace surface. But such a hyperspace surface needs a hyperspace in which it is embedded. Nevertheless, this idea, neatly formulated mathematically, would not correlate with practical physics because time and each path are inter-dependent over speed. However, since physics is the science of moving matter, models such as theories of relativity are absurd with movements independent of time.

Infinity is not sensually penetrable. That's why we fail here. Even the Nobel laureate for physics, Brian Schmidt failed on 15th Sept. in 2015 on Australian television on the question of a child, because, as Stephen Crothers reported, he regarded the infinity as a mathematical value [3.14]. The child's question was: How can something as immense as the universe grow larger? On the one hand, the cosmos is regarded as infinite, on the other hand Brian Schmidt and two other colleagues received the coveted Nobel Prize in 2011 "*for the discovery of the accelerating expansion of the Universe through observations of distant supernovae*". Obviously, the Nobel Prize Committee, which is based at the Royal Swedish Academy of Sciences, is not competent in this question, even though Sweden has produced two outstanding plasma cosmologists, Oskar Klein and Hannes Alfvén. The latter won the 1970 Nobel Prize in Physics. Today they defame their cosmos model as pseudoscience [3.15]. But when they fight something, they confess fear, fear of the truth. We will discuss why they are afraid of this in Chapter 6 The Macrocosm.

3.5.1 How to overcome the Dilemma with Infinity?

System theory has an answer to that. It offers an open system with just such subsystems with feedback. The term system theory first appeared in 1949 in Karl Küpfmüller [3.16] in connection with communications engineering. The idea is to describe a concrete actual situation as a system, so to speak, as a black box by its external behavior completely by a few system characteristics, without knowing anything about the internal processes. The method introduced by Küpfmüller had a high potential of generalization for other scientific and technical facts.

The inadequacy of the knowledge of a starting point in control engineering and in other technical applications, as well as in cosmology, aroused the need for a theory of system behavior in any past and finally led to a new theoretical conception of the system description, the description of the system state. The concept of the system state, which has always been fundamental to classical physics, has also generally become the main concept of a general system conception and a modern system theory [3.17].

Subsequently, the ideas of system theory are briefly outlined and their advantages compared to the standard model of cosmology are highlighted. Starting from Hamilton's system of motion equations, we can interpret the moments p and the places q as states of the mechanical system of the motion of mass points or stars. Thus, instead of the differential equations for the coordinates and impulses, we obtain the simple form

$$\dot{z}_i = dz_i/dt = \partial H/\partial z_{l+i} = H_i(z_1, z_2 \ldots z_{2l})$$

As a result, instead of a dynamic system, we obtain a system of static states at specific times, which is comparable to a movie clip. The q_i are replaced by the z_i numbered from 1 to l and the pi by the z_i from l + 1 to $2l$. The $2l$-tuple z $(z_1, z_2 \ldots z_{2l})$ then designates the state of the Hamilton system at time t and the set Z is called state space. The state $z(t) = z$ is a $2l$-tuple of variables from which not only the spatial position of the mechanical system at time t can be taken, but also all its future positions in space, if there are no collisions. For this, the knowledge of the past of the system is not required. As far as it plays into the future, it will be sufficiently taken into account by the current state. Thus, the system's past has become completely uninteresting, since it can only be reconstructed in time-invariant systems anyway. The standard model of cosmology is based on time reversal. It projects the image of the actual state into the past by means of the spectral redshift, which is inadmissible interpreted as escape velocity. Here we have the old idea of the Laplace demon in a new guise. The Standard Model does not consider the natural evolution of stars from their formation to their death. I. Prigogine first dealt with evolutionary systems in 1972 [3.01] by extending thermodynamics to open systems beyond thermodynamic equilibrium. These systems are open to energy and mass flow. However evolution systems far from thermal equilibrium, as well as the cosmos are not time-invariant but open to the flow of matter, as everybody can learn at the game of life by J. H. Conway very easily. An example has been given in section 3.4.1 Causality and Coincidence. To describe the cosmos, we need more variables than the p and q for the movement of all stars, namely their state of development, for example, based on the spectral class from the Hertzsprung-Russel diagram, which maps the stellar evolution. Again, we are faced with the problem of a necessary data reduction. As time and place in a dynamical system depend on each other over speed, which leads to the failure of the theory of relativity in a space-time continuum, another thought is added through the automaton theory. This theory describes time-discrete systems, (how should time also be quantified in cosmic dimensions?) Where inputs from a set ***X*** are answered with outputs of set ***Y***, which are not real numbers, as must act in the case of *p* and *q*. Here, unlike the Hamiltonian system, a distinction is made between forced variables (the inputs x from a set ***X***,), the free variables (the states *z* from a set ***Z***), and the observable variables (outputs *y* from a set ***Y***). At the same time two functions are defined.

f: ***Z* × *X*** →***Z***, *f(z,x)* = *z'* as a transfer function from one state to another and

g: ***Z* × *X***→ ***Y***, *g(z,x)* = *y* as a result function of the dynamic system

> You have to read the above expressions as follows: From the set of all combinations of the elements of the sets of states ***Z*** and the inputs ***X***, which are imaged to the set **Z** of states, we image a function *f* with the independent elements *z* and *x*, onto the Elements *z'*. This bypasses the problem of time, which the theory of relativity has. We consider static states and no dynamic that is in terms of our earthly history to the history of the cosmos a snapshot. In the case of the function g, this is the present image of the visible cosmos.

The difference to a Hamilton system is that the Hamilton system is closed and there is no result function. The states are assumed to be observable. The given locations *q* can be regarded as inputs *x*. The advantage of our system definition over the Hamilton system is that it does not require source freedom of the vector field, which explicitly allows the appearance and disappearance of mass points. The game of life can take place. Stars may arise and pass away. The question of how the system gets from one state to another can be determined individually. Neither translational nor rotational movements are mandatory. A successful example of this approach is the simulation of a galaxy performed by A. Peratt [3.18]. When we make the output of one system to the input of another system, systems can be coupled. The consideration of subsystems makes the modeling of natural processes much easier.

In the generalization of this approach, interactions between two and more systems can also be defined. Even networks of systems are in principle writable. Systems theory is still a young science, yet it has already achieved success across disciplines. Today, relational databases allow us to treat practically multidimensional state spaces. An example of this is the Sloan Digital Sky Survey project, where the observable actual state of the sky is documented. It provides the set Y of the current system Cosmos.

3.6 The Doppler Effect

The Doppler effect (rarely Doppler-Fizeau effect) is the temporal compression or elongation of a signal with changes in the distance between transmitter and receiver during the duration of the signal. It was predicted by Christian Doppler in 1842 for the light. For the sound waves, Christoph B. Ballot proved this effect in 1845. He positioned several trumpeters both on a moving railway train as well as next to the railway line. In passing each one of them should blow a G and the other should determine the heard pitch. It was not until 1865 that William Huggins discovered the predicted spectroscopic Doppler shift in the light of stars. He showed that Sirius is steadily moving away from us.

3.6.1 The Doppler effect in Sound Propagation

We know the Doppler effect of a car driving past us, approaching us with a higher tone than it leaves us again. The driver in the car does not notice. He always hears the engine with the same frequency as long as he does not change his speed. Obviously, the appearance of the Doppler effect depends entirely on the point of view of the observer to the emitter of the vibration. So it has nothing to do with the physics of the vibration source, but is exclusively a function of the observation site.

Let's take a numerical example: The propagation velocity of the sound in the air is c_0 = 340 m/s at normal temperature. The sound energy is calculated

$$E=\oint J\cdot dS \qquad (3.08)$$

J is the sound power produced by the engine. $J = p_{eff} \cdot v_{eff} \cdot \cos Ø$. The circulating integral expresses that energy must be integrated over the entire path *S*. The sound pressure p_{eff} is the pressure that the observer feels on his ears and v_{eff} is the speed observed at the roadside, i.e. the acoustic beam that hits the observer. Since the engine is not accelerated, the sound intensity is constant and thus the energy dissipation for the entire system also, so we do not have to consider the energy conditions here. Thus, the effective sound pressure at the effective change in speed, which the observer detects at the roadside, must change so that the energy remains constant. In fact, it is found that low frequencies at the same energy have a higher sound pressure and are not perceived only by the ears.

The speed of sound is calculated as $c = \lambda \cdot f$ The car is driven at a speed of 57.6 km / h, which the driver of the car reads from his speedometer. That corresponds to 16m / s. The driver continues to see on his tachometer 3000 revolutions / min. In a four-stroke engine, an ignition occurs every two revolutions, which the observer can hear at the roadside. So that's 1500 / 60s = 25 Hz. The wavelength is then $\lambda = c / f$ = 340/25 m = 13.6m. Now there is an observer at the roadside who sees the car approaching. The hears due to the sound velocity of the sound source a frequency which results from $f = (c + v) / \lambda$ = (340 + 16) / 13.6 Hz = 26.2 Hz and when the car leaves again, he hears one Frequency $f = (c - v) / \lambda$ = (340 - 16) / 13.6 Hz = 23.8 Hz. However, the motor itself did not change its rotation while passing the observer. The driver can confirm this in the car. So the sound intensity *J* in equation (3.08) has to be the same for both observers, because it is the same car as used for the effective speed in the above calculation example.

The Doppler effect is not the only effect that has its cause in the relationship of the observer to the observed object, and thus is based on a hallucination.

- We remember that the roadside observer first saw the car quite small, and it got bigger as it approached the observation point, and then it seemed to get smaller after leaving the observation point. But our experience has taught us that it isn't a physical effect that may inflate and shrink the car, but that has something to do with the geometrical law of perspective.

- Another example of the different view between stationary and moving observers is the special theory of relativity, which predicts a shortening of an object in the direction of motion at high speeds and a time dilation. Physically, these effects can not be explained. These are simply the laws of projection that lie behind them and are found in the Lorentz transformations identified as imaging processes.

3.6.2 The Spectroscopic Doppler Effect

Let us now turn to the Doppler effect of the light. The Doppler effect of light is a typical example of an inductive conclusion. In the year 1865, the spectroscopic Doppler effect of William Huggins was found. The sound propagates in air at a temperature of 20° C at a speed of 343 m/s and is based on the vectorially addition of the speeds of sound source and observer. In our sound experiment, all key measurements were known. This is not the case with the Doppler effect of light. If we want inductively to infer from the propagation of sound to the conditions in the light, we should transfer the ambient conditions to the new medium. So here we also have to demand that the speeds of light source and light add up, as in the above example, which should not exist, according to relativity theory because of the required constancy of the speed of light. Otherwise we could not perceive a Doppler effect. We have no observer on the light source, which can reliably transmit frequency and velocity when it is a cosmic object. While we had the sound wavelength as a link between the two observers, now that we observe the wavelength, we must now consider the frequency of the light to be constant for both observers, since the conservation of energy requires for the closed system (3.08)

$$E = h \cdot \nu = const. \qquad (3.09)$$

with the light frequency v and Planck's constant *h* for electrons. Also, there must be a medium (force field) that transmits light at a speed that depends on the material properties of the medium, and as mentioned previously, the speeds of light and light source with respect to the observer must be able to be added vectorially. This requirement contradicts Einstein's hypothesis that the speed of light is constant across all reference systems. There are two serious objections to his hypothesis.

- The speed of light, like the speed of sound, is a material constant $c^2 = 1/\varepsilon \cdot \mu$, where ε describes the electrical and μ the magnetic material properties of the cosmic plasma. If there is perhaps only one atom per cm^3 in the cosmos, then it acts as a mass with its surrounding field and the probability that it is neutral is rather small, because the reticular structure of the cosmical luminous matter indicates intense force fields. Einstein has regarded the speed of light as constant, ignoring the entire optics, which contradicts the vector addition of light speed and airspeed of the observer.
- The Sagnac effect, whose technical application is the gyroscope, clearly shows that the speed of light depending on the movement of the light source and the rotation of the earth really add up vectorially. If that were not the case, there would be no Doppler effect. When should the point be reached that the normal vector addition of velocities from one reference system to another transforms into Einstein's speed hypothesis of the constancy of the speed of light in all frames of reference? There is no limit point.

Einstein's demand for constancy of the light speed comes from the Lorentz equations, which require a constant projection center and which is physically not justified. That's one of Einsteins blurred traces. (Look at his quote in 2.2) His quotes are of great sincerity and should be taken seriously if you want to get close to him in order to truly understand failure of his theories.

Back to the Doppler effect of the light. In our experience, the blue shift of the light can be interpreted as an approximation and the redshift as a removal of the light source. The shift can be detected particularly well on the spectral lines of the light spectra of the stars. The delineation of this theorem requires asking, what speed limit of the light source is presumable. In sound we have the phenomenon of supersonic velocity for a sound source. Is there maybe an over-light speed for a light source? Light sources have mass. Because masses can not reach the speed of light, the superluminal speed is certainly not possible. Nevertheless, in the standard theory of cosmology, a phase of superluminal velocity is assumed to justify the expansions of the visible cosmos. In chapter 7.3 Cosmic Velocities we will deal with this topic.

3.7 Modeling, Experimental Design and Measurement

When we hear the term *model*, we remember model railroad or airplane model making, or other models that we used to play with as children and that led us into the real world of adult life in our dreams. These models were imperfect and our imagination has equipped them with the desired features. The same is true with models in science. A theoretical model is intended to depict the essential properties of a natural process without actually achieving reality, as Immanuel Kant has already stated. This means that the model is similar to, but not identical to, the modeled natural process. The more knowledge about the natural process is present, the more we can approach the model of nature. The type of model is determined by my watching aspect. It defines the methods used for modeling. In the past, scientific modeling was a mysterious creative process whose course remained hidden. Today, even the scientific work is examined and algorithms are derived, as the causal relationships in nature is traced. For the detection of causal relationships, an example should be given here.

In general, modeling begins with the recognition of a problem, a knowledge deficit. Through inductive analogies from existing knowledge, the second step leads to a scientific conjecture, a hypothesis. In the next step, we must work out a method to disprove this assumption, as Karl Popper demands. However, it is the common practice to want to prove the presumption. To do this, we must identify causes and effects and determine functional relationships. If a fact fits the presumption, the proof seems to be provided. But there may be several facts to match a guess.

For example, it was discovered the cosmic background radiation and this was considered proof of the Big Bang. However, each body radiates according to its temperature a continuous Planck spectrum whose maximum is the temperature of the body. This is the same with the cosmos as it is with the sun. How should then the temperature radiation of the cosmos prove the Big Bang?

Then the test conditions have to be set. Finally, you can start planning the experiment. The method of statistical experimental design takes this circumstance into account. With that, it is possible systematically to refine a model with reasonable experimental effort, without much initial knowledge about the nature of the functional relationship. Two measurement points already provide a linear relationship and three a squaring relationship of a parable. If you have more than one independent parameter of the experiment, you can vary them simultaneously. Usually you work with over-determined polynomials from more measurement points than are necessary for the polynomial, which is why the experimental design is a statistical technique with a quality criterion for the model obtained. Thus you obtain both a validity range for the model and an error probability for an intermediate value derived from it. (You can approximate every curve in a given range by a polynomial.) The series expansion of all basic mathematical functions makes this possible. All our computers can only perform logical functions and addition, but algorithms from sequences of these basic operations allow the construction of complex arithmetic operations.

Statistical experimental design is rejected as a method by Theoretical Physicists because it would provide no knowledge about the obtained model parameters. These parameters (natural constants) would be purely statistically verified. But how else do they want to verify these parameters? Newton based his law on Kepler's data. He has determined no validity limits, which are missing in many other established laws of nature too, which is why it is falsely assumed that these equations have universal validity. For example, a singularity has not been observed in nature. How do you want to measure that a thing? Our measuring systems can only measure finite quantities. Consequently, our conclusions can only move in finite areas. A physicist can therefore never speak of infinity and zero, but only of excessively large or very small. In the future, there may be the possibility of extending the current measurement limits.

After the data acquisition by means of measurements, the statistical processing of the data follows. There are a large number of analysis and test methods to separate the measurement signals from interfering signals. At the end of such a modeling process, the theoretical evaluation is carried out to ensure that the result is consistent in the existing knowledge. But that's not always the case. Then the hypothesis was wrong and the problem analysis starts all over again. Natural science models are therefore subject to refinement and change through the progress of knowledge. Even if models are more or less fragmentary, they can be useful if they can explain a fact. Thus, the Bohr atomic model is useful because it can explain the spectral lines. However, the standard model of particle physics can not explain the radioactive decay of isotopes, which is why it is useless.

An extreme position represents the astrophysicist Mario Livio of Johns Hopkins University in Baltimore and author of the book *Is God a Mathematician?*, with the thesis :

> *»Kindly, nature is governed by universal laws and not by field-forest-and-meadows rules with limited range.«* [3.19]

This requires trust in God and contradicts any progress of knowledge.

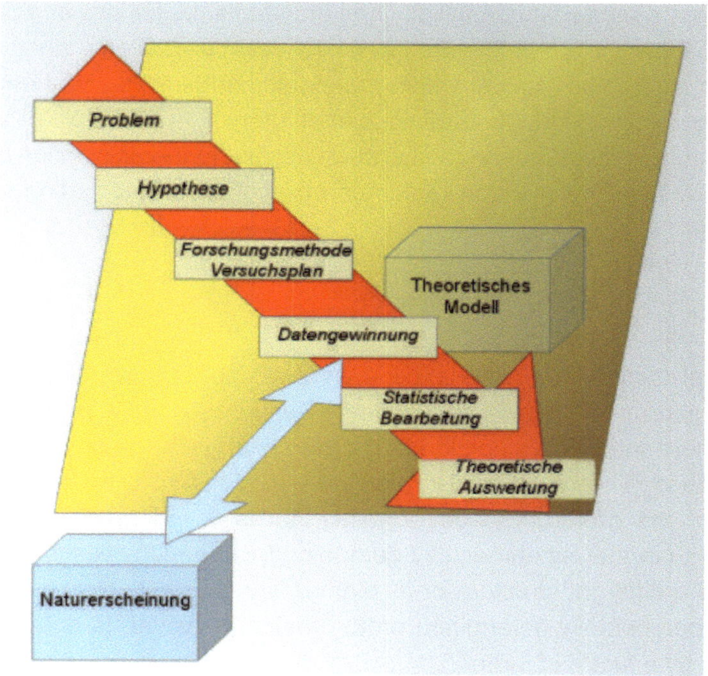

Figure 3.6 Development process of a theoretical model

As an engineer, I would not want to trust this statement without verification, and this review is likely to be difficult. There are certainly some basis laws of general validity that we have already discussed in Chapter 2, with all the others I would say that we do not know most of the borders. What we consider to be universally valid laws only applies within the limits of our sensory or technical perception. As we move beyond old boundaries due to improved measurement techniques, we can unexpectedly enter areas where we need to review our thinking models. A very concrete case has come through space technology since the middle of the last century. We left the interplanetary perspective of a Galileo and come in with Voyager1 and our telescopes into an intergalactic angle of view.

Moreover, a model whose validity has proven its usefulness in a particular area leads beyond this area to misleading conclusions, since each model can only be verified within the experimental limits. Mathematics does not follow nature, theoreticians often deny it. An example of this is the formula for the electron radius when applied to the calculation of the proton radius. The electron radius can be calculated according to the equation.

$$r_e = \frac{e^2}{4\pi \cdot \epsilon_0 \cdot m_e \cdot c^2} \tag{3.10}$$

It did not originate from the experimental design, but from the analogy of charged macroscopic hollow spheres and describes a charged hollow sphere with the distribution of the elementary charge on the spherical surface. The result can be verified experimentally. If you use the mass of the proton, you get a value that is 3 orders of magnitude smaller than the experimentally determined value. Obviously, the proton has a different structure or its elementary charge is bigger or both. The experimentally determined proton radius, however, is only slightly smaller by a factor of three [5.04].

It is therefore always necessary to be careful when drawing conclusions from equations that go into unknown territory. Physics as a science draws its conclusions from the observation of nature.

Conversely, it is also true that we can not draw knowledge from unobservable items. Whoever crosses this border, goes into the land of fantasy. This is not so bad as long as you remain aware of it and do not try to draw more and more conclusions from hypotheses. The scientific modeling process requires that theory and experiment are very closely interlinked. So it is the models that shape our idea of the world itself. But our idea is not identical with the world, unlike Schopenhauer's intention to persuade us. Therefore, we on the way to better recognize the world, are sometimes subject to the mistakes of deceptions and self-deceptions of which magicians can live.

The specialization in natural sciences in the 20th century led to theoretical physics being almost completely detached from experimental physics. If you look in the textbooks of theoretical physics; this gives the impression that theoretical physics is a special field of mathematics. Therefore, the method of modeling here is another. In the theory-related areas experimental design, data acquisition and statistical analysis only have an alibi function. Paul Dirac is said to have aptly formulated this: »*If any experiment contradicts a beautiful idea, let us forget the experiment.*« [3.20] In another source, this quote is attributed to Albert Einstein. Obviously, it was not conceived as a joke but as a guide to action. The proponents of theoretical modeling argue: '*In theoretical modeling, physical processes of nature are analyzed and formulated mathematically using the laws of science. In this way, the model structures and, as far as possible, the model parameters can be determined via the internal mechanism of action. The mathematical models are therefore scientifically justified*'. It is assumed that a model developed in this way would have universal validity. The model parameters are interpreted as natural constants. This is undoubtedly a sham, as they do not specify the description of the method of analysis. The disadvantages of the theoretical modeling are therefore to be seen in the unreliability of insufficient knowledge of the natural phenomenon and the high modeling effort in complex natural phenomena. We have seen the role of the observer's philosophical attitude and knowledge in modeling nature.

Ultimately, such modeling without experimental practice leads one to increasingly withdraw to areas completely inaccessible to observation and experiment, as elucidated in the book *The Black Hole War* by Leonard Susskind, where the question is discussed whether the information falling into a black hole, is preserved or not. This question is asked without the question of the information carrier. The information as well as the energy can not be separated from its carrier, as we will see in section 6.4 The Battle for Black Holes. For physics theorists, a black hole is as real as the nine-quarters platform in London's King's Cross for Harry Potter fans. We, who are not blessed with magic, orient us on measurable things.

The most important prerequisite for modeling is therefore the measurement process. In physics, we should only talk about measurable things. Unfortunately, modern physics no longer adheres to this principle. They speaks of black holes and dark energy in the cosmos, quarks and gravitons in particle physics and other fantasy entities that have never been measured. This is akin to medieval geography that has filled the oceans with mythical creatures.

If we want to measure something, we have to compare quantities. The classical mechanics essentially came out with three properties to be measured. These were the length, the mass and the time and for this we need the corresponding units of measurement, that are comparison or reference dimensions, with which we compare all occurring lengths, masses and time intervals. It goes without saying that these reference measures for a public space have been made binding by a political authority. It hides behind it neither a scientific law nor a deeper mystery. The fact that this was a political decision was forgotten only, since these definitions are already so much older than today's usual political decisions are valid. Let us remember: we owe our metrical system of measurement to the French Revolution. Before, each kingdom had its own system of measurement. In Germany it was introduced only during the Napoleonic foreign rule. The calculation in the metric system with the base 10 is only that old. The question of why a natural constant has exactly the value and no other, is probably primarily because we count people with the ten fingers. If we had kept our fingers crossed, the natural constants would have a different value based on 8. As you know, mathematics is logic based on definitions and agreements. In principle, it is completely immaterial how the agreements are made, it is crucial that these agreements are valid for very long periods and are accepted everywhere.

A physical measuring consists of a numerical value and a unit of measure or a decimal multiple of this unit. A total of 4 base units are sufficient to measure all electromechanical processes. Almost all over the world today, the MKSA system (**M**eters, **K**ilograms, **S**econds, **A**mperes) applies with exception of three countries. That units of measurement can change at political borders is well known. For example, in the US, the inch is used instead of the centimeter with the conversion factor 2.54. However, it is often overlooked that physical influences can influence our measurement system. It must be taken to ensure that the fixed unit dimensions do not change even in use in different situations and find the inevitable changes due consideration. Since the reference dimensions have physical properties, they are certainly subject to change, especially when they come into another force field or are subjected to acceleration. We will come back to that later.

It is a known fact that energy supply in the form of heat leads to the expansion of bodies, which is why the original meter has been dispensed with today and since 1983 the base unit meter has been defined as follows: **1 meter is the distance the light travels in a vacuum in 1 / 299.792.458 seconds.** However, the vacuum is not defined.

A pendulum clock slows down when heated, because the pendulum lengthens. It is also influenced by the gravitational field. The ratio of length to time interval, however, remains constant, which is why one has defined the unit of measurement meters over the speed of light and the time interval second. The second is based on the 9,192,631,770 times the period of the radiation corresponding to the transition between the two hyper-fine structure levels of the ground state of atoms of the ^{133}Cs nucleate. The original kilogram of 1884 is still valid for the unit of mass.

3.8 The Principle of Relativity

Finally, when we talk about the observer, we also have to talk about the principle of relativity. Today, this basic principle of physics is formulated in this way:

»*The principle of relativity is the premise that the equations describing the laws of physics have the same form in all permissible reference systems.*«

Is that true?

Relativity expresses that things or properties are related to each other. But that means there are things too that are not related to each other. We have already stated that you also summarize things or properties that have no relationship with each other, otherwise you could not recognize the relations, because our system of knowledge is fixed on the difference. An example of this is the mathematical concept of space. Because the dimensions are perpendicular to each other, they have no relationship with each other. This means that you can change the value of one dimension regardless of the value of all other dimensions. A relation between two things or properties is an ordered pair. For example, a function is an ordered relation between a subset x of the set ***X*** and a subset y of ***Y***. Each measure represents a relation between a number from the set of real numbers and the set of physical units. However, what makes the principle of relativity worth noting is the question: If two observers observe a phenomenon from different points of view, then do they see the same or do they make different observations, or in other words is the observation dependent on the observer's viewpoint or not? Because relations are ordered pairs, we must therefore determine whether the observer or the appearance has the primacy. Reasonably enough, no one will deny that a physical phenomenon is independent of the observer, unless one argues with Schrödinger's cat, which is supposed to be dead or alive depending on the behavior of the observer. One of the most important relations of physics is that of motion and rest. Movement is always measured against a resting reference point whose characteristic is inertia. Einstein called this an inertial system.

The larger mass is always the resting system compared to the smaller moving system. So the Sun is the dormant system to Earth, but the Sun is moving in the Milky Way system. The center of the Milky Way is the dormant system with respect to the planetary system and the Sun. The Milky Way, in turn, is part of a much larger system whose motion we have not yet discovered. That, however, in all these systems the speed of light should be an absolute constant, is a false assumption of Einstein, on which he has built his theory of relativity. The movement of the celestial bodies is relatively small compared to the speed of light because of their large mass, yet, as the optics shows, it is not negligible, otherwise there would be no optical Doppler effect. We can summarize: The principle of relativity determines, according to the largest mass and its power to be considered, what the resting center is. The equations of physics are aligned with that. We can summarize and reformulate the principle of relativity:

The principle of relativity determines, according to the largest mass to be considered, what the resting center is. The equations of physics are aligned with that.

So, unlike Einstein claims, it makes a difference whether the station or the train is moving, or whether the sun is orbiting the earth, or the earth is orbiting the sun. You see that Einstein's relativity abolishes the church's dispute with Galileo over celestial mechanics.

In this chapter we have dealt with the errors and misunderstandings of the observer as dialogue partner of nature. Let us now turn to the other dialogue partner, nature or the world, as the physicists prefer. To interpret natures signals correctly is one of the most important tasks of science. In doing so we depict nature in our minds by means of language (mathematics). Mapping processes are clear, but not irreversibly clear, as experience has taught us. We go into the realm of so-called virtual reality with images. Doing physics there is pure magic, such as Lewis Caroll shows us in Alice's Wonderland. The consequences of this will be discussed in the following chapters, but for now let's take a look at what the matter of the world is all about. The concept of matter is currently used very misleadingly, which is why we have to take it more strictly. We want to discuss with that in the next chapter.

Figure 3.7: Cheshire cat from Alice in Wonderland - John Tenniel (1865)

4 For Terminology of Physics

»When the terms confused, the world is in disarray« - Confucius

Physics is one of the most difficult to understand natural sciences. That was not always so. It is due to the increasing contradictions that modern physics has created with relativity and quantum mechanics. They apologize that these disciplines have not been sufficiently explained.

Is that just the explanation? As a scientist, you should also be able to explain what you understood. We remember Kant's division of the terms into *phenomena* and *noumena*. Natural sciences, including physics, deal with phenomena that can be sensed and try to explain them. In contrast, *noumena* are pure intellectual inventions from the realm of fantasy. The effort to reconcile faith with science causes the boundaries between the two conceptual worlds to be blurred.

The gateway to such blurring are the idealizations of scientific phenomena. An example is the mass point. Everyone sees an object in a certain place that fills a certain volume, and can feel its weight when lifted. In order to be able to mathematically formulate the laws of motion in physics, the phenomenon of 'volume' is reduced to zero in the imagination and the phenomenon of 'heaviness' is united in one point, the focus. This abstraction is called a mass point, knowing that there is no such thing in nature, because each mass has a volume consisting of a mass of points. But the statements made about this model are sufficiently precise for the classic applications of solid body mechanics. But even with liquids the equations are no longer usable and with gases each atom or molecule is an individual mass point with its own movement.

The mathematical movement of a point is described by a transformation. If I only describe the result of a point transformation, I have a Euclidean transformation ahead of me. This transformation maintains form. One says it is form invariant. If I take into account the speed with which the transformation takes place in relation to the speed of light, I am dealing with the Lorentz transformation. The Lorentz transformation is not shape invariant. However, it is not suitable for mass points because it does not take the masses into account in the transformation, it only takes place in the spirit. When the mind works on matter, it is called magic. In both cases it is not a real physical transformation of a real object. For a real physical transformation, an additional energetic consideration has to be made. You know the story of the Prophet and Mountain movement. "If the prophet cannot come to the mountain, the mountain must come to the prophet." If they ignore the mass of prophet and mountain, the theoreticians call it relativity. I think now you're feeling a bit foolish. This feeling is completely fine at this point, as you will see later.

Because physics works with idealizations, the consideration of errors is a very important thing. Errors describe deviations between theory and practice. However, where you have no sensual or metro-logical experience, you can no longer recognize errors. The expansion of natural laws beyond the world of experience is also a gateway for faith. Especially when it comes to orders of magnitude that go beyond our imagination into both the macro and the micro world. These include borderline questions such as: Is the world and the being finite or infinite, is there a beginning or not?

Even at the time of the birth of Buddhism, priests and monks were arguing about it. Some said that the world is infinite and the others that it was born out of nothing. In *Udāna VI, Part 5* of the *Pali Canon of Theravada Buddhism*, it is said: »Surely, some mendicants and priests cling to such questions, and in these they go under, having failed to dive into nirvana.« Also for science these questions are meaningless because they can not be answered scientifically. Only charlatans give an answer and they pretend to know exactly what happened a second after the big bang, and those who believe them, are themselves to blame. It's not as bad as the promises of big winnings on dubious racketeers, but "The big bang is just marketing," says Nobel Laureate Robert Laughlin in 1998 about the misbelief in a Theory of Everything [4.01].

In 1814, Pierre-Simon Laplace formulated the following sentence in the foreword to the *Essai philosophique sur les probabilités* [4.02], which on the one hand became a guideline for the physicists of later generations and on the other a stumbling block:

> *»We may regard the present state of the universe as the effect of its past and the cause of its future. An intellect which at a certain moment would know all forces that set nature in motion, and all positions of all items of which nature is composed, if this intellect were also vast enough to submit these data to analysis, it would embrace in a single formula the movements of the greatest bodies of the universe and those of the tiniest atom; for such an intellect nothing would be uncertain and the future just like the past would be present before its eyes.«*

This strictly deterministic world view is called the *Laplace's demon*. It follows the idea that God created the universe with His laws in the way a watchmaker would build the perfect clock. Once created and brought to the correct initial state, the universe relentlessly runs according to the will of Divine Providence. The opponents of this view, inspired by the evidence of Henri Poincaré, that already three bodies tend to chaotic behavior, argued as follows:

> *»...But even if the laws of nature had no more secrets to tell us, we would only approximate the initial conditions. If this allows us to specify the following states with the same approximation, we say that the behavior was predicted to follow laws. But that's not always the case: It may happen that small differences in the initial conditions have great effects in the results, a prediction is impossible then, and we have a random phenomenon.«* [4.03]

However, when Immanuel Velikovsky's book *Welten im Zusammenstoß* [4.04] appeared in the middle of the 20th century, it triggered a worldwide scandal, which Velikovsky, inspired by none other than Albert Einstein, documented in the book *Sternengucker und Totengräber* [4.05]. Einstein studied very intensively on Velikovsky's book during the last months of his life, as the exchange of letters between the two shows [4.06]. Velikovsky's merit is that he placed the electrical nature of the cosmos at the center of his considerations. However, he drew his findings not from physical considerations but from the myths of ancient peoples. He claimed that in prehistoric times humanity witnessed cosmic catastrophes, and that these catastrophes were no more than ten thousand years ago. There have undoubtedly been catastrophes in the history of mankind. But earthly catastrophes perhaps triggered by sun storms, may have already deeply affected the consciousness of ancient people that they appeared as cosmic disasters mirrored in the human imagination, lacking a clear idea of the depth of the cosmic space.

Inspired by the Laplace demon, many theoretical physicists still dream of finding the TOE, the so-called holy grail of theoretical physics. For this they are looking for a way to synthesize the theory of relativity, a deterministic world image, with quantum theory, a statistical world image, into a theory of quantum gravity.

> *»The holy grail of modern physics is the theory of quantum gravity. It is the search for a view of the universe, the two seemingly opposing pillars of modern science: Einstein's theory of general relativity, that of phenomena on large scales like planets, the solar system and galaxies and the quantum theory that deals with the world of the very small - molecules, atoms and electrons.« from the back cover of the book The Three Roads to Quantum Gravity* [4.07]

The specially developed string theory has already failed and the second attempt, loop quantum gravity [4.08], will also fail. At the same time, electrical forces do not seem to be associated with the gravitational forces, but that are not the forces that are the problem but the fundamentally different watching aspects of the two theories. And Lee Smolin spoke of three roads in his book [4.07] The third road (it is a narrow pathway) he left open to those who do not believe in relativity and quantum theory. Here we will follow this third path, strictly adhering to established philosophical principles, observing what exists within and outside of our consciousness.

4.1 The 4 Phases of Matter

The division of matter into four manifestations (Greece: phasis = appearance) dates back to the Greek philosophers five hundred years before our era. Heraclitus of Ephesus saw the primordial substance in the ever-changing fire, driven by inner contrasts. The Far Eastern thinkers of that time had developed similar ideas. Smoke sacrifice and the burning of sacrificial donations are still used in the Hindu, Buddhist tradition to contact the gods and make their prayers heard. The custom of lighting a candle for someone on a Christian altar may trace back to the same roots.

The term *phase* for the physical state of different domains of matter is used in thermodynamics. Classical physics uses the term aggregate state. Although the conceptual separation made here is not physically justified, it does show the different aspects of consideration. Thermodynamics uses the statistical approach, whereas classical physics prefers the causally determined way of looking at things.

> *»This world order, the same for all beings, has created no god and no man, but it was everlasting, and is and will be its eternally living fire, glaring to moderation and modestly extinguishing.«* - *Heraclitus*

By Heraclitus of Ephesus you can already find the division of the original material in four different phases, which also brought Pythagoras a little later with him to Italy.

- Earth - today we call it the solid state of matter,
- Water - today we understand all liquids under this category
- Air - today we see in it the gaseous aggregate state and
- Fire - today we call glowing gases the fourth phase, the plasma that fills 99% of the cosmos.

"*Prometheus stole the fire from the sky and brought it to man, for which he was forged as a punishment from Zeus to the Caucasus and where an eagle ate daily from his liver*," the legend says. Did the fire come from the gods or did people consider the fire of an aurora borealis itself to be a divine manifestation? It was once sacred to mankind. Today's astrophysicists believe in an other god. For them, the plasma is meaningless, because it is virtually neutral for them. They do not consider that this quasi-neutrality is only considered from the outside for a closed system. But we are part of the cosmos and experience it as an open, energy-flooded system.

Aristotle added a fifth phase, the quintessence, the ether. The Five Element Doctrine is found in the Far East in Taoism and Buddhism. However, there the division is a bit different: earth, wood, metal, fire and water are these five elements.

Also at the time when Heraclitus developed his idea of the basic material, Leukipp first talked about the divisibility of substances and he suggested that divisibility could not be continued indefinitely. There must be a limit to divisibility for substances. These particles are indivisible, and in ancient Greek means 'indivisible' atomos, i.e. atom. Each object is then represented by the mass of its inseparable particles. By this it is completely irrelevant which form has the original body or these particles have. But if matter consists of particles of different kinds, there must be forces between the particles that guarantee cohesion, so that these substances can form bodies and occupy a volume in space. The question is therefore, do the forces belong to the mass-carrying atoms or do they form a separate force field and what are the reasons of these forces? Are they possibly particles or effects? It is undisputed that we can not see forces. They are recognizable and measurable only by their effects. What the relationship between mass and force field is, we will examine in this chapter.

This force field was still called ether at the beginning of the twentieth century and by means of interferometric measurements according to Michelson and Morley was searched for, and since the result was much lower than expected, it was rejected as non-existent in order not to endanger Einstein's theory of relativity. It was not recognized that the experimental setup in a cellar was completely unsuitable to answer this question, but only the attempt of the french physicist George Sagnac brought a partial answer, namely that the speed of light source and light add up or subtract like ordinary speeds. Einstein, on the other hand, claimed that the speed of light was constant in all inertial systems. This will be discussed in detail in the section 4.5 The Puzzle about Light and its Speed.

Today, a lot of dark matter is talked about. You might think that the first three aggregate states of matter are meant, which in fact do not emit light. However, here is meant a matter form whose kind should be unknown. It is an empty Kantian term that will express the lack of understanding of certain cosmic phenomena. The phases of matter have received very little attention in the past. I put the phases at the beginning of my world view, since all our information comes from phase boundaries. All our senses are designed to capture the difference.

4.2 The Forces

> *»Modern physics distinguishes four types of fundamental forces, gravitation and electric force, as the external forces and forces inside the atomic nucleus, dividing them into weak and strong forces. Each of the four basic forces of nature comes about through the exchange of elementary particles in virtual states. There is a presumption that these are only different forms of the same force — to prove this, but has not yet been successful. The power comes through exchange of bosons between fermions. That is, the effect of the force is that*
>
> - *pulling fermions towards each other (as if they were connected by a rubber band)*
> - *or keep them at a distance from each other (as if they were connected by a spiral spring, which can not be compressed arbitrarily far).«* [4.09]

What does tell us the above explanation? - Nothing! - It just raises new questions.- What are *fermions* and what are *bosons*? But we want to put these questions back. We return to this in chapter 5 The Microcosm

We feel forces. For example, if we hold two magnets in our hands, if we want to get up, if we want to move an object, if we rub a piece of cloth, in short, if we want to do a job. We remember: In classical physics, *force* is the reason of an action that can deform a locked body and accelerate a moving body. The acceleration is a change in the speed of the body. It is always a directed size, since bodies move in a distinct direction. The force practically inherits the property 'direction' from the acceleration. We speak of vectors. As a force $\boldsymbol{F}$ is a vector. Force vectors always have a direction and a value measured in units 'Newton'. When talking about the basic forces, we should therefore assume that we should distinguish forces according to their basic directions and not according to the places where they have been found [4.10]. In three-dimensional space, there are three basic directions that can be used to guide all directions can generate. Consequently, there should only be this classification. On the other hand, metro-logical, forces can only be distinguished by their amount. There is little point in distinguishing them according to the manner of their appearance outside and inside the atomic nucleus. So far we have been talking about the application of a force to the movement of bodies, so we have to clarify that a bit. A body is described by its volume and mass. However, bodies of the same volume may contain different masses. Consequently, in physics, bodies are represented by their masses. Forces are therefore the product of mass and acceleration. When a car accelerates, it changes its speed. Braking is a negative speed change. Because speed is the ratio of a change of location within a time interval, we can describe the force as

$$\boldsymbol{F} = m \cdot d\boldsymbol{v}/dt \quad \text{or} \quad \boldsymbol{F} = m\, d^2\boldsymbol{s}/d^2t. \qquad (4.01)$$

We read the equation as force is the product of mass and velocity change in a time interval. The latter is also called acceleration. The gravitational force is therefore the mass of a body multiplied by the gravitational acceleration downwards to the earth. The acceleration is also the second derivative of the way after the time. I beg the pardon to my readers for following to use so many equations. (If you do believe what I said, then you can just skip the equations. However, if you do not believe me and that is your right, you will have to fight through it. It's like in the days of the Reformation when the Bible has been translated by Martin Luther from Latin into German. Here the fraud is based on the ignorance of the language of mathematics among the public.) The product $m \cdot dv$ is called impulse change dp. A pulse produces on another mass an effect via power coupling. This follows directly from causality. Impulses are therefore retained. Equation (4.01) tells us that forces are always connected to masses. You can not bind forces to a space, as Einstein did in his General Theory of Relativity. It is unclear how a physical force on a mental concept such as space, should exert a force to retroactively effect a change in reality. Einstein was aware of what he did when he published the General Theory of Relativity. He wrote to Paul Ehrenfest in 1917:

> *»Ich habe auch wieder etwas verbrochen in der Gravitationstheorie, was mich ein wenig in Gefahr setzt, in einem Tollhaus interniert zu werden.« Translation: I've given an offense again in the theory of gravitation, which puts me in some danger of being interned in a madhouse.«*

but he was not aware of the consequences of this joke at that time, because later he said:

> *»If I had known the consequences, I would have become a watchmaker.«*

Because this idea was tucked on by a scholar namely George Lemaître and thus the alleged academic joke became serious.

Remember: Mathematical space is a concept. The physical counterpart is the volume. Moreover, in a space with four dimensions, in which one is supposed to be the time, we can not have speed, since time t and distance s in such a space should change independently of each other, and consequently speeds and accelerations make no sense.

Now we have to ask the question: How do we distinguish the four types of basic forces? As the force is a vector, forces can only be distinguished by direction and value in units of Newton. Metro-logically, we can not distinguish the basic forces at all. Even St. Hawking admits that this is an arbitrary man-made grading [4.11].

$$1\text{Newton} = 1\text{kg}\cdot\text{m/s}^2$$

There is no point in distinguishing forces by species if they can not be differentiated metro-logically. We do not have to prove this fact, as it results by definition.

4.3 The Energy Equations and their Significance for Physics

Summing up the impulse over all velocity changes gives the energy consisting of kinetic and potential energy. The integral is reminiscent of an 'S', which means that you sum up the patches of small speed changes multiplied by the corresponding amounts of impulses.

$$E = \int p dv = m \int v\, dv = \frac{1}{2} \cdot m \cdot v^2 + E_{pot} \qquad (4.02)$$

The energy flow is t h e movement form of matter. The kick-off is given by the impulse.

Energy is simply what our modern society keeps on running. In every-day life, we encounter energy in various manifestations, which can be derived from the mass as electrical, chemical and nuclear energy as well as mechanical (kinetic and potential), magnetic, and thermal energy. Its characteristic feature is its ability to transform from one form into the other, depending on which properties of the atomic shell or the nucleus are being addressed. This is expressed in the sentence of energy preservation.

The equivalence between masses and energy is ascribed to Einstein:

$$E \Rightarrow c^2 \cdot m \quad \text{for} \quad m \to 0 \qquad (4.03)$$

Although in the concepts of his theory of relativity there is no meaning whatsoever, because relativity only works without mass. In the formula (4.03) mass and speed of light are reversed in row to emphasize the importance of mass as a carrier of energy, while speed acts as a proportionality factor and the speed of light represents an upper bound on smaller and ever smaller masses.

In fact, as early as 1880, in the context of electrodynamics was known about the proportionality between energy and mass. Einstein's merit is the postulation of the proportionality factor c^2, the square of the speed of light, the constancy of which, on closer examination, can not be sustained, otherwise the laws of optics would be overridden, as we will discuss in the section 4.5 The Puzzle about Light and its Speed. Equation (4.03) with its both limits states too that the energy-mass relationship is preserved. Energy and mass are two properties of matter that seem to be inter-convertible, but do not disappear. This is a fallacy, as thermodynamics teaches. Looking at the four states of aggregation, it can be seen that the energy content of matter increases from the solid state to the plasma state, although the mass density decreases and consequently the atoms can absorb more kinetic energy. Like force too, you can not see energy, which is why some people, even physicists, think that energy is not material. But this idea is likely to evaporate at the latest when you see a victim of a lightning strike. "Lichtenberg Figures" on the skin, higher-grade burns, atrial fibrillation and cardiac arrest are not the result of an idealistic concept, but material violence. That is the consequence of unclear concept formation. We still find the equalization of the meaning of matter and mass. Philosophers and physicists do not understand each other.

Confusion is always caused by the two expressions: $E = m \cdot c^2$ and $E = 1/2\ m \cdot v^2$. Which is the right one? Letting v go against c and m against zero, there is no apparent reason why the energy should double. According to the rules of integral calculus, the factor 1/2 is necessary. The equation becomes understandable only when interpreted as a contact potential. Since the speed of light is the highest speed we know, equation (4.03) represents a limit to the energetic potential of a small elementary particle. with a mass near zero. The kinetic energy, on the other hand, is the usable energy flow. The flow is maintained until the energy potentials between the source and the sink have equalized. If no additional energy is added to the system, then exactly half of the energy from the source will flow out and fill the sink. That's exactly $E_{kin} = 1/2\ m \cdot v^2$.

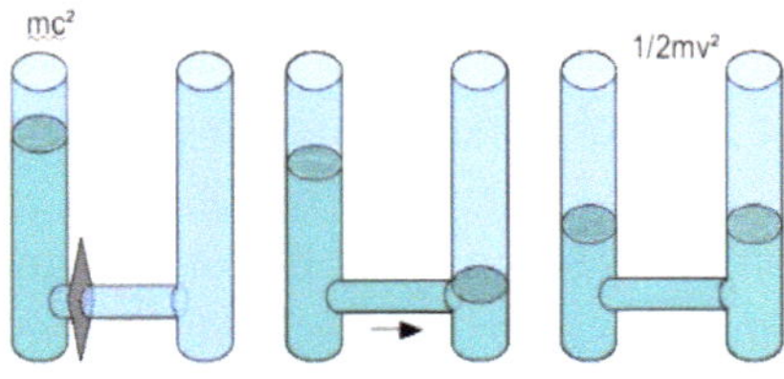

Figure 4.1: The energy flow in a closed system

We have already learned another energy equation under (3.6). This equation contains a frequency. The frequency can be modulated. Thus, the flow of energy becomes an information carrier. It could be called kinetic information. The clock pulse of a ticker is the simplest information. Every radiation carries information. This information is used by the spectroscopist to learn about the radiation source. The broadcasting industry imprints its own information on the radiation source. In contrast, there is the information that can be read from immovable structures, such as a book of illustrations or such as phonograms. It could be called potential information in analogy to potential energy. This is the third property of matter besides mass and energy. But physicists are only interested in very simple modulations. From the simplest form, the clock pulse, they derive the time.

According to equations (4.02) and (4.03), energy as well as information are bound onto carriers like impulses or masses, depending on whether they are moving or stationary. An energy change in mechanics at a certain speed is always associated with a momentum change. You also can write (4.02) in form of

$$\Delta E = \boldsymbol{v} \cdot \Delta \boldsymbol{p} = \boldsymbol{p}/m \cdot \Delta \boldsymbol{p} \rightarrow E = (1/2 \cdot m)\, p \qquad (4.04)$$

In thermodynamics, an energy change at a certain temperature level is always associated with an 'entropy' change, since here the energy source appears to be entropy, formerly called heat. It has the unit Joule/Kelvin. It is the part of the energy that is technically no longer re-convertible and thus no longer usable. This, however, contradicts the idea that the entropy S is identical to the information, as information is a structural property of mass or energy.

$$\Delta E = T \cdot \Delta S \qquad (4.05)$$

The term entropy comes from the Greek and means immutability. It has experienced a number of interpretations in the past. We will discuss this in more detail in section 4.7. The Entropy, in everyday life has the meaning of heat. Because heat is an expression of the disordered speed of the atoms of a substance, entropy in chemistry is an expression of disorder in an aggregate state. In computer science, entropy describes the permeability of a news channel. This is because the modulation of the energy flow carries the information. Both amplitude and frequency modulation are possible. We remember: 1 bit is the information content contained in a choice of two equally probable possibilities. In the computer, the information is realized by turning on and off the electrical power. This is an electrical energy change in the sense of an amplitude modulation in its strongest form. An electrical energy change is always associated with a charge change at a certain voltage (a potential difference). Because here the charge change, which is called electrical current in electrical engineering, is the energy carrier, the following applies:

$$\Delta E = \mathrm{F} \cdot \Delta Q \rightarrow W = U \cdot I \qquad (4.06)$$

Where F is the electric field intensity and Q is the charge and U is the voltage and I is the electrical current. This is the field of electrodynamics. Another energy change results from the gas kinematics, where at a certain pressure, the volume change is the energy carrier.

$$\Delta E = p \cdot \Delta V \qquad (4.07)$$

Where there p mean the pressure of the gas and V is the volume. Pressure and momentum in their meaning have to be kept apart. So you can transfer the idea to the chemical energy transport and for the transport of nuclear energy.

Equation (4.03) also states that mass can not arise from nothing and that energy seems not simply be lost without creating mass. On the other hand, the mass of a body seems to decrease when it gives off energy in the form of radiation. A mass defect Δm causes a change in energy ΔE, which usually manifests itself in the form of thermal or other radiation.

If we look at the equations (4.04) to (4.07), we see that they all describe a particular branch of physics. In order to describe nature, we must study the totality of their movements and their interactions. The available energy results from the sum of the kinetic energies of a given mass released by the individual potential differences.

$$\Delta E = v_m \cdot \Delta p + T_m \cdot \Delta S + \Phi_m \cdot \Delta Q + p_m \cdot \Delta V + \ldots \qquad (4.08)$$

If any equation has the potential for a theory of everything, then it is the energy equation that retains its validity in the specific form over all phases of matter.

We see that *energy* is the term that connects all disciplines of physics and that the mass is in any case its bearer. This encourages you to continue along the path you have taken and to take a closer look at the components of matter.

4.4 The Mass

The mass seems so familiar to us that we do not need to talk about it. Because matter has a particle structure, the mass is a quantitative property. In our language, a *mass* is an uncountable set of things, objects, or appearances. The mass in the physical sense is a bit more complicated than it may seem at first sight. We know the mass with the unit of measure kg only as comparison with the Primary kilogram, which is stored in Paris. To explain mass via the energy equation leads to a circular conclusion. For understanding the mass we need a mass spectrometer.

We have obtained the most information about mass from mass spectroscopy. The fundamental equation of mass spectroscopy says that the mass flow corresponds to the product of charge and magnetic field strength. In other words, the gravitational force can not be distinguished from the electromagnetic force, as stated above, because Coulomb's force equation differs from Newton's force equation in the structure 'only' by a factor of 2.3×10^{39}. One can get an idea of this factor by comparing the diameter of an elementary particle with the length of a cosmic super-cluster full of galaxies. That corresponds to 39 orders of magnitude. In other words, a kg of pure protons exerted a force on an electron cloud of 0.54 g, comparable to the gravity of 2 billion solar masses. The Milky Way is estimated at 1.5 trillion solar masses. After all, that would be comparable to the gravity of 1.3% of the Milky Way.

The mass is defined as a comparison with a reference body that rests in the earth's field. So far, we only differentiate masses by comparison with each other. Some people may still remember the old beam scales with their weights. On the other hand, the mass in the mass spectrometer is measured by the charge amount of the ions impinging on different points of the detector.

Einstein postulated that the inert mass was identical to the heavy mass. This sentence has become known as the equivalence principle. According to Wikipedia, the equivalence principle is listed in two forms: *»According to the weak principle of equivalence of all the properties of a body, alone its mass determines(i.e. the measure of its inertia) which in a given homogeneous gravitational field gravity acts on the body. Its other properties such as chemical composition, size, shape etc. have no influence. According to the strong equivalence principle, gravitational and inertial forces on small distance and time scales are equivalent in the sense that they can not be distinguished in their effects either with mechanical nor any other observations. From the strong equivalence principle follows the weak; whether this applies the other way around may also depend on the precise wording and is not yet conclusively clarified.«* Given the fact that the mass is always related to a reference body, the distinction between inert and heavy mass is irrelevant, and the weak equivalence principle says nothing other than that a quiescent mass does not change in the force field. Things are different when we look at mass on the basis of its electromotive properties in motion. The equivalence principle then no longer applies, as we shall see.

The inertia of the mass we feel, for example, when we want to set our car in motion. In contrast, the heavy mass is determined by the gravitational mass attraction. Now, however, it can be stated that in this example the mass is displaced once perpendicular to the gravitational field and in the other case the mass will be heaved against the gravitational force. The difference here is that the heavy mass rests in the field and the inertial mass experiences an additional acceleration in a certain direction counteracted by its inertia. Now everyone has the experience that he can indeed push away his car but can not lift. This is not up for discussion. The difference is that the inertial mass needs an impulse to move, which in accordance with the relationship 'action equals reaction', immediately must propagate the reaction in the body, while the Gravitation on each atom constantly attacks and acts on the surface without any movement. The gravity is permanent, the impulse has only a limited time. The heavy mass stores potential energy. We now want to see what happens when an impulse acts on a potential.

4.4.1 The Mass from a Macroscopic Point of View

The starting point is the highest possible potential, which is described by equation (4.03). We superseded in equations (4.03) the mass by the impulse, divided by the speed at which it impact, we get:

$$E = \frac{p}{v} \cdot c^2 \qquad (4.09)$$

Also it applies to the flow of energy: (We write ∂E instead of ΔE, if we mean partial differentiation, ie the gradient of an area in the direction of each variable.)

$$\partial E = F \cdot \partial x = \frac{\partial x \cdot \partial p}{\partial t} = v \cdot \partial p \qquad (4.10)$$

Equations (4.09) and (4.10) show:

$$E \cdot \partial E = c^2 \cdot p \partial p \qquad (4.11)$$

Now we will integrate equation (4.11):

$$\int E \partial E = \int c^2 \cdot p \, \partial p \qquad (4.12)$$

The integration provides:

$$E^2 = c^2 \cdot p^2 + E_0{}^2 \qquad (4.13)$$

where $E_o{}^2$ is the integration constant and is called rest energy. The rest energy must therefore contain the heavy mass. The factor 1/2 occurs here on both sides of (4.13) and can therefore be removed by multiplying the equation by 2. Equation (4.09) transformed also yields:

$$c \cdot p = E \cdot \frac{v}{c} \tag{4.14}$$

Equation (4.14) in (4.13) gives:

$$E^2 = E^2 \frac{v^2}{c^2} + E_0{}^2 \tag{4.15}$$

Equation (4.15) solved for E yields:

$$E = \frac{E_0}{\sqrt{1 - v^2/c^2}} = \gamma\, E_0 \tag{4.16}$$

As it is a quadratic equation, there is even a second solution that has the same value only with a negative sign. This says that the energy used is given back in reverse. The factor

$$\gamma = \frac{1}{\sqrt{1 - v^2/c^2}} \tag{4.17}$$

is the same one we know from the Lorentz transformation (3.03). It is referred to as Lorentz factor. It was introduced by Joseph Larmor in 1890 and was named after the mathematician and physicist Hendrik Lorentz. Substituting the mass-energy relationship $E_0 = m_0 \cdot c^2$ and $E = m \cdot c^2$, equation (4.16) yields:

$$m = \frac{m_0}{\sqrt{1 - v^2/c^2}} = \gamma \cdot m_0 \tag{4.18}$$

How is the result to be interpreted? We have given an impulse to a potential and consequently receive the inertial mass m. With m_0 we have the resting mass in front of us, before it has received the impulse. The gravitational potential is still there. A moving mass increases with increasing speed? But mass is an uncountable set of particles. Do particles arise from nowhere? That's magic, the entrance to metaphysics.

Because it is a moving mass, the mass increase must be a force that simulates a mass increase. Suppose an observer could move in the same direction as our observed mass at the same speed v . He would carry his steelyard with him and see no change in the mass, because the number of elementary particles has not changed, which also says our newly formulated relativity principle from section 3.8 The Principle of Relativity. In order to determine the speed, it needs a fixed reference system. Physical reference systems are always bound to mass and thus to a force field, which is why dormant always refers to the bigger mass, in contrast to Einstein, who speaks of an inertial system but does not define it mathematically. The resting mass is therefore not equal to the moving inert mass, but appears from the perspective of the dormant system by a factor of γ greater, although it has not changed from the perspective of the moving system. Einstein's Relativity is therefore a matter of consideration, not of physics.

Einstein's principle of equivalence assumes that the resting mass equals the inert mass. However equation (4.18) shows that as the root approaches zero with increasing velocity *v*, the increase in mass of a moving particle seems to be a direct consequence of the preservation of mass and energy, without the requirement of a relativity principle or constancy of the speed of light, as Albert Einstein asserted in his paper [4.12] from 1905. The mass in equation (4.18), which can not increase arbitrarily, is therefore erroneously called the relativistic mass. It should be better called inertial mass, as it is speed-dependent, while the heavy mass is identical to the rest mass. Equation (4.17), solved by the velocity *v*, yields:

$$v = c\sqrt{1-\left(\frac{E_0}{E}\right)^2} \qquad (4.19)$$

Equation (4.19) tells us that mass particles, in contrast to photons, can never reach the speed of light, since the right-hand expression supplies the speed of light *c* only for $E_0 = 0$, which would only be possible in the case of $m_0 = 0$.

Contrary to Einstein's postulate we distinguish here strictly between inert and heavy mass. Since we have seen that the theory of relativity does not describe reality, we use the terms *heavy mass* and *rest mass* synonymously, as well as *inert mass* and *relativistic mass*.

Only the fact that the speed of light propagation is finite limits the kinetic energy for a given mass. This in turn causes the supply of energy to increase the inertia of the mass, further reducing its speed.

We now want to know what the reason is for the alleged growth of the mass. We remember that it is said that gravitational and electrical forces should act outside the atomic nucleus. The laws of Coulomb and Newton are structurally alike. This brings us to the idea of considering the mass in the light of electrodynamics.

4.4.2 The Electrodynamic Mass of an Electron

If we have derived the inert mass above according to the laws of classical mechanics, we now want to derive them from the laws of electrodynamics. For this we use equation (4.20).

$$\Delta E = \mathrm{F} \cdot \Delta Q \rightarrow W = U \cdot I \qquad (4.20)$$

We want to determine the mass of a magnetic field. It is obvious to everyone that in order to drive a mill wheel, a flow of water is needed that provides the kinetic energy for the drive and that can be used to calculate the flow rate of water as mass per unit of time. The same must be possible for an electric motor by determining the magnetic flux that generates the magnetic field. This must correspond to one mass per time unit. We specify the task to the extent that we want to know how much mass an electron increases when it moves at a certain speed. Now it gets a bit demanding. But without this calculation, you certainly will not believe that the magnetic field has a mass that increases with increasing velocity of the electrons.

First of all, we imagine a wire, which is traversed by a current and forms a magnetic field. In what follows, we will consider the reflections of Paul

Marmet from his essay *The Fundamental Character of Relativistic Mass and Magnetic Fields* [4.13]:

He proceeded as follows:

- First, he calculated the magnetic field of a single moving electron as a function of its velocity.
- In the second step, he calculates the magnetic energy that induces the electron through its motion.
- From this he has determined the total mass of the magnetic field as a function of the velocity. Using equation (4.18) he calculates the inertial mass of the electron and subtracts it from the total mass of the magnetic field. This completes our task.
- Now we still have to check, whether this mass increase is identical to the mass increase of inertial mass. For this we calculate the inertial mass of the electron according to equation (4.18) and subtract the magnetic mass increase from the total mass of the magnetic field. We expect to receive the heavy rest mass of the electron. If that is the case, we have shown that the mass increase of the magnetic field is exactly equal to the mass increase of the inertial mass in the case of its motion.

We know that the *electric current I* is defined as a number of individual electric charges (*e*-) going through one wire cross-section per second. Since the electric charge is quantized, in the case of a single electron, it is impossible to calculate a minute change in charge from a single electron. Electrons can not produce an uninterrupted flow of electric charge. This is particularly obvious when the number of electrons is close to one. In order to measure a current of one ampere at an ammeter, must by the ammeter $N_{(1\ \text{Ampere})}$ @ 6.25×10^{18} electrons per second flow. The electron current *I* is defined as the passage of one coulomb of the electric charge *Q* per second. We have:

$$I = \frac{dQ}{dt} = \frac{d(N_e)}{dt} \tag{4.21}$$

Because the electron charge is quantized, the arrival of a single new electron corresponds to the emergence of a new charge *dQ*. Therefore $dQ = d(N_{e\text{-}})$. The electron velocity *v* is defined as the distance *dx* that travels the electric charge along the wire per second. The velocity of the electron flow is constant. Equations (4.21) will be like this:

$$I = \frac{d(N_e)}{dt} = \frac{d(N_e)\cdot v}{dx} \tag{4.22}$$

From Maxwell's equations we can derive the scalar form of the Biot-Savart equation. This equation describes the magnetic flux density around the conductor, which more accurately describes the change in energy from (4.06) in terms of geometry:

$$dB = \frac{\mu_0 \cdot I}{4\pi^2} \cdot \sin(\theta)\, dx \tag{4.23}$$

Substituting equation (4.22) into (4.23) yields:

$$dB = \frac{\mu_0 \cdot v}{4\pi^2} \cdot \sin(\theta)\, d(N_e) \tag{4.24}$$

We see that along the wire, the Biot-Savart equation provides an increase in the magnetic field *dB*, generated at a distance *r*, as a function of the velocity of an elementary charge *dQ*, which is distributed along the length of the wire. When we calculate the magnetic field generated by a single (isotropic) electric charge, there is no longer a linear distribution of charges. Therefore, equation (4.15) must be changed to account for the change in the geometry of the electron source.

Quantization of charges: Equation (4.24) gives the component of the magnetic field described with the flux density ***B*** in the direction of *Q*, generated by an electric charge consisting of *N* electrons distributed along the wire. At Maxwell's time, it was still unknown that the electric charge was determined by discrete electrons, which was not noticeable given the amount of electrons in the conductor. Now we want reduce the number of electrons to 1. Equation (4.24) must be checked to see if it is still valid because of the disappearance of the linear distribution of the electric charge.

Now that we consider the magnetic field for a single electron, additional adjustments are needed. In the case of a single electron, we can not determine a directional angle Θ between continuous charge distribution and magnetic field. Since the axis of electric charge distribution no longer exists, we must find a new geometry. Since we now have an isotropic electric field around a single electron, let's assume that the magnetic field that is generated is also isotropic. Equation (4.24) becomes:

$$dB_i = \frac{N \cdot \mu_0 \cdot e \cdot v}{4\pi^2 \cdot r^2} d(N_e) \qquad (4.25)$$

Here, $d\boldsymbol{B}_i$ is the calculated magnetic field for a single isotropic electric charge with no axial charge distribution. You can not measure such a weak magnetic field, yet there is no reason to doubt the validity of (4.25).

The induced magnetic energy around a single moving electron.

From the induced magnetic field we now want to calculate the magnetic energy around a single electron. The magnetic energy density u_m, which is defined as the magnetic energy U_m per unit volume V, is given by the ratio (4.26):

$$u_m = \frac{U_m}{V} = \frac{B^2}{2\mu_0} \qquad (4.26)$$

Because the magnetic field $d\boldsymbol{B}_i$ around a single electron, as calculated in Equation (4.25), is the only magnetic field of interest, integration omitted and we simplify the notation below by replacing the magnetic field $d\boldsymbol{B}_i$ in Equation (4.25) with the simple symbol B. Here it is tacitly assumed that the electron has no dipole field, but a rotation field like the conductor. Jan de Climont [4.14] has experimentally proved that this condition is fulfilled. To obtain the magnetic field around a single electron (N = 1), equations (4.25) and (4.26), we calculate the magnetic energy dU_m within a volume dV. Then is:

$$dU_m = K \cdot \frac{v^2}{r^4} dV \qquad (4.27)$$

with

$$K = \frac{\mu_0 \cdot e^2}{2(4\pi)^2} \qquad (4.28)$$

Equation (4.27) gives us the magnetic energy dU_{m} in volume dV around a single electron.

Total Mass of magnetic energy in a single moving electron

Equation (4.27) gives us the total energy of the induced magnetic field around a moving electron in volume dV. Now let us calculate the total mass M of this magnetic field surrounding the moving electron using (4.26). We know from (4.03) that the proportionality factor between energy and mass in this dimension is c^2. Since Equation (4.27) gives the energy per unit volume dV, it must be divided by c^2 to obtain the mass of the magnetic field. We get the mass density to:

$$dM = \frac{dU_m}{c^2} = K\frac{v^2}{c^2 \cdot r^4}dV = \frac{\mu_0 \cdot e^2 \cdot v^2}{2(4\pi)^2 \cdot c^2 \cdot r^4}dV \qquad (4.29)$$

We want to calculate the total magnetic energy (and mass) in the unlimited volume around a single electron. Because the integration of Equation (4.29) contains a singularity, we must find the appropriate integration bounds for the magnetic mass. In electromagnetic theory, the magnetic field expands to a moving electron (Eq. (4.26)) to infinity. Therefore, the total mass of the magnetic field surrounding the electron must be integrated throughout the three-dimensional space, to infinity. We found above that the distribution of the magnetic energy is isotropic, since the electron has a spherical geometry. To integrate the total mass of the moving electron's magnetic field, we apply the volume integral of a sphere on which we use a variable radial density, which was calculated in equation (4.29).

In Figure 4.2 we see that the differential surface element of the surface of a sphere in a distance r equals a thin rectangle with a side length of $r \cdot d\theta$ along the meridians, multiplied by the element of longitude $d\varphi$ of the circle $2\pi r \cdot \sin(\theta)$.

The total volume of an ordinary range is given by the double integral:

$$V = 2\pi \int_0^{\pi} \sin(\theta)\, d\theta \int_{r_{min}}^{r_{max}} r^2 dr \qquad (4.30)$$

In Equation (4.30), the volume of a region between radius zero and radius r_{max} is, as expected: $V = 4\pi r^3/3$.

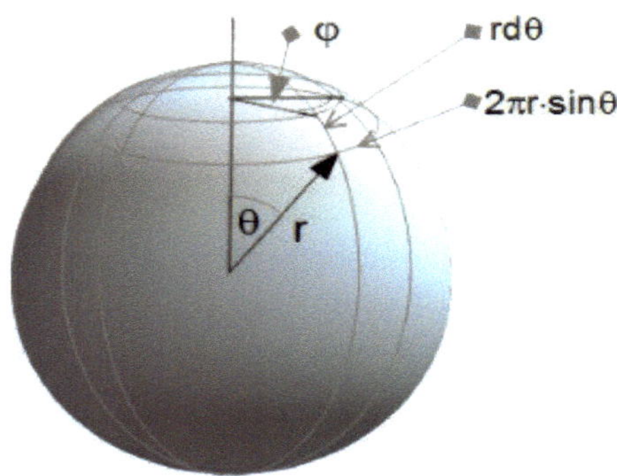

Figure 4.2: illustrates the parameters used in Equation (4.21).

In the case of the magnetic energy that expands to infinity, the upper integration limit for the radii in Equation (4.30) is infinite. We must then consider that the density of the magnetic mass is variable and decreases with $1/r^4$. To calculate the total mass of electromagnetic energy within a volume given by Equation (4.25), we must integrate the density distribution in Equation (4.29) with the volume in Equation (4.30). This double integral is integrated as follows:

$$M = \frac{\mu_0 \cdot e^2 \cdot v^2}{2(4\pi)^2 \cdot c^2} \cdot 2\pi \int_0^{\pi} \sin(\theta)\, d\theta \int_{r_e}^{\infty} \frac{r^2}{r^4} dr \qquad (4.31)$$

Equation (4.31) can be written:

$$M = \frac{\mu_0 \cdot e^2 \cdot v^2}{16\pi \cdot c^2} \cdot \int_0^{\pi} \sin(\theta)\, d\theta \int_{r_e}^{\infty} \frac{1}{r^2} dr \qquad (4.32)$$

The sin(θ) integrated gives the cos(θ). By substituting the integration limits we obtain from the first integral 2. The total mass of the magnetic energy in Equation (4.32) remains limited even if the upper integration limit is infinite. However, we note that Equation (4.32) gives an unlimited mass (magnetic energy) at r = 0. We know that an unlimited mass is physically unrealistic. There is obviously a physical limitation that must now be considered.

Magnetic fields can be measured well over long distances, but a measurement in the center of the electron is virtually impossible. At $r = 0$, equation (4.32) shows a pole, which would require an infinite amount of energy to produce a magnetic field to the center of the electron. Therefore, the resting energy of the electron of 511 keV gives the information about how close its field can reach to the geometric center of the electron. As a result, because of the limited energy of the electron (511 keV), the interior of the electron is practically field-free. Due to the fact that the electron energy (511 keV) is limited, a minimum radius is expected called the classical electron radius r_e. There within the classical electron radius r_e, no electromagnetic energy can exist as calculated by the Biot-Savart equation.

We want to integrate the magnetic energy from the well-known classical electron radius r_e to infinity where an electric field can exist. From equation (4.32) we have the magnetic mass:

$$M_{r_e \to \infty} = \frac{\mu_0 \cdot e^2 \cdot v^2}{8\pi \cdot c^2} \cdot \int_{r_e}^{\infty} \frac{1}{r^2} dr \tag{4.33}$$

After integration we obtain for the entire mass of the magnetic field of an electron moving at velocity v:

$$M_{ges} = \frac{\mu_0 \cdot e^2 \cdot v^2}{8\pi \cdot c^2} \cdot \frac{1}{r_e} \tag{4.34}$$

The increase in electron mass due to their velocity from the point of view of the dormant observer

We want to compare the magnetic mass of a moving electron, as given in equation (4.33), with its inertial mass. When we applied the principle of mass-energy preservation above, we found that the inertial mass m_v of a moving particle is given by the same relation as in Einstein's theory of relativity but without Lorentz transformation. The mass of a moving electron is given by the relation (4.18)

$$m_v = \gamma \cdot M_e \qquad (4.35)$$

From equation (4.35) the increase of the mass due to the velocity is:

$$\Delta m = M_e(\gamma - 1) \qquad (4.36)$$

In mathematics, we can show that a series expansion of g yields:

$$\gamma = 1 + \frac{1}{2} \cdot \frac{v^2}{c^2} + \frac{3}{8} \cdot \frac{v^4}{c^4} + \frac{5}{16} \cdot \frac{v^6}{c^6} + \ldots \qquad (4.37)$$

Equations (4.35), (4.36) and (4.37) show:

$$\Delta m = M_e(\gamma - 1) = \frac{M_e}{2} \cdot \frac{v^2}{c^2} + \ldots \qquad (4.38)$$

In equation (4.37), the expression second order $(v/c)^4$ is extremely small when v is much smaller than the speed of light. The ionization energy of hydrogen is 13.6 eV, which corresponds to 9.6×10^{-19} kg m²/s². Divided by the electron mass and pulled the root, we get 1026.6 km/s as the rate of detachment of the electron from the proton. It follows that $(v/c)4 = 1.3 \times 10^{-10}$ is. The term $(v/c)^4$ and other higher order terms are therefore negligible with respect to the first term. It can be shown that these higher order terms are caused by the energy needed to accelerate the mass increase due to the previous expression $(v/c)^2$.

The magnetic mass in comparison with the inertial mass.

Let us compare the increase of the magnetic mass calculated in equation (4.38) above with the increase of the electron mass using formula (4.35). In fact, we test whether the inertial mass is identical to the magnetic mass. Equations (4.34) and (4.38) give:

$$\frac{\mu_0 \cdot e^2 \cdot v^2}{8\pi \cdot c^2} \cdot \frac{1}{r_e} \Rightarrow \frac{M_e \cdot v^2}{2c^2} \qquad (4.39)$$

We note in Equation (4.39) that both phenomena (magnetic energy and inertial energy) produce an increase in mass that is proportional to (v/ $c)^2$. This means that the magnetic energy around individual electrons increases with the square of the electron velocity, just like the increase in the inertial mass. For the complete identity of the magnetic and inert masses, we then have to compare only the proportionality factors between these two phenomena. A resolution of equation (4.39) to M_e yields:

$$M_e = \frac{\mu_0 \cdot e^2}{4\pi \cdot r_e} \qquad (4.40)$$

The electron mass is given as $M_e = 9{,}109383 \times 10^{-31}$ kg. The remaining constants are of the highest known accuracy

- The magnetic field constant of the vacuum is $m_0 = 4\pi \times 10^{-7}$ N/A^2 . 1 N =1 kg·m/s^2
- The electron charge is: $e^- = 1{,}602176565 \times 10^{-19}$ As
- The classical electron radius from the tables is: $r_e = 2{,}8179403 \times 10^{-15}$ m

Using the values in Equation (4.40) yields:

$$M_e = \frac{4\pi \cdot 10^{-7} N}{4\pi \cdot A^2} \cdot \frac{1{,}602176565^2 \times 10^{-38} A^2 s^2}{2{,}8179403 \times 10^{-15} m} \qquad (4.41)$$

and further:

$$M_e = \frac{1{,}6021765652^2 \times 10^{-45}\, kg\, m\, A^2 s^2}{2{,}8179403 \times 10^{-15}\, m\, A^2 s^2} \tag{4.42}$$

The numerical calculation gives exactly $9{,}109383 \times 10^{-31}$ kg.

This shows that the magnetic mass of the moving electron within the experimental accuracy is identical to the increase in the inertial mass for any velocity of the particle. We find that the two numerical values in Equation (4.39) are physically identical. In both cases, there is an identical increase in inertia due to the velocity with respect to the electron mass at rest.

The increase in inertial mass is actually nothing more than the resistance of the generated magnetic field against the speed of the electron.

The so-called 'relativistic increase in mass' has its real cause in the velocity of the electron, which induces a magnetic field dependent on its velocity, as shown by the Biot-Savart equation. The faster the electron moves in a force field, the greater is the resistance presented to it, which is represented as the magnetic 'mass' in the equations, which is why the prognosis that the Lorentz force would disappear at almost the speed of light is obviously wrong and thus Electrodynamics, contrary to the doctrine, is Galileo-invariant, if the above conclusions are taken into account. Or in other words; since mass is defined as the innumerable quantity of particles, it is not possible to generate additional electrons, but the magnetic force, which generates additional inertia, creates the illusion of an increase in mass. At the same time, we see that at the level of elementary particles, the boundaries between force and mass dissolve. Another conclusion is that the speed of light must decrease due to the increase in the permeability μ. This causes the light to shift red.

A conclusion from sections 4.4.1 and 4.4.2 is that the force field can not be seen detached from the mass, but is an integral part of the mass.

So far we have looked at the motion of a free electron and calculated its mass increase. But we know that electrons are only released above the ionization energy. But what happens to the electrons below the ionization energy when they are bound in the atom and the atom as a whole is accelerated? We will explore this question in the next subsection.

4.4.3 The Accelerated Hydrogen Atom

In the previous section, we have seen how a free electron appears to increase in mass when accelerated. Now let us turn to the question of what happens to the atom when it is accelerated?

Because the equation (4.18) for the inertial mass is not subject to any restrictions, it must therefore also apply to the individual atom. To avoid unnecessarily complicating matters, consider a simple hydrogen atom consisting of a proton and an electron, and rely on the Bohr atomic model, as it is simple enough for plausible reasoning, though it has the known weaknesses and has been replaced by the quantum mechanical model. The Bohr model is static, meaning that the atom is at rest. So far, hardly anyone except Paul Marmet has thought about what would happen to the atom if it were shot from a 'nuclear gun' in a vacuum. The connection between electron and proton, after all, is not a rigid connection, but elastic due to the prevailing field. However, the H atom must not exceed the peel rate of 1026 km/s to be ionized. We have seen that the mass of the accelerated electron increases. Imagine the atom as a rotating wheel whose circumference is formed by the electron orbit. As the mass of the rotating wheel increases, its rotational speed decreases because of the preservation of the angular momentum. Considering the Coulomb forces, the radius of the electron orbit becomes larger as the electron velocity slows down. Consequently, the Bohr radius of the electron orbit increases as the entire atom moves faster. Even when energy is supplied in the form of heat, a mass expands. That's a familiar fact. Maxwell and Boltzmann found a relationship between the temperature and the velocity of the atoms. Then, why should the mass contract one-dimensionally when accelerated, like Einstein claims? We have explained in section 3.1 that the Lorentz transformation is a really illusion, a virtual effect that only the resting observer would sees.

The change in the Bohr radius is a likely cause of mass expansion when atoms receive kinetic energy. Consequently, no spatial direction with respect to expansion is preferred, which is in contrast to Einstein's theory, who assumes a one-dimensional shortening.

This mechanism is calculated in detail in the article by Paul Marmet *The Natural Physical Length Contraction Due to Kinetic Energy* [4.15]. Here I give just a summary of his theses:

This increase in the Bohr radius also causes a shift in the atomic energy levels, and thus shifts the wavelength of the emitted radiation into the red region. As a result, a moving atomic clock is now running at a slower rate. Even pendulum clocks run slower with extended pendulum. Quite naturally, we see how the increase in size of the radius of the electron orbit and the enlargement of the macroscopic mass come about.

Although Einstein's theory of relativity predicts a virtual (as we know now) length contraction in the direction of motion, it does not explain how a mass can contract physically or why this phenomenon is not reversible when the mass in the moving frame is accelerated back to the original frame. Einstein's length contraction meant in the real world that the electron radius would have to be smaller in the direction of motion. However, quantum mechanics shows that such a contraction of the electron radius should increase the atomic energy level, which would be indicated by a blue shift of the spectral lines. The consequence of Einstein's predictions contradicts the observed facts, which show that at high speed the atomic energy levels become smaller and atomic clocks run slower.

Of course, if the mass expands as a result of the increase in energy during acceleration and the density of the accelerated body decreases, this also applies to our standard meter and our standard second will be slightly longer compared to the standard second in the stationary frame of reference.

Because of the increase in mass with speed, the size of the units of measurement changes in moving coordinate systems. That also affects length and time. This change in the size of the units of measure must be taken into account in physical transformations.

We have to realize that a physical measure is the product of the number of units of measure multiplied by the corresponding unit of measure. Now, this everywhere on earth constant unit of measure is no longer constant as soon as we leave the earth, but in its size depends on the reference frame to which it belongs. This is because the unit of measure is represented by a material body subject to the physical laws of its environment. We have a stationary earthbound frame and a moving reference system. Then we can describe a mass by means of two indices, using m_s for stationary and m_v for moving. In addition, each of the two reference systems [s] and [v] represents an observer with his specific units of measure, which can observe both the stationary and the moving mass. This gives four possibilities: m_s[s] and m_v[s] for the stationary observer and m_s[v] and m_v[v] for the moving observer. Then relations between these four masses must be found in order to be able to draw the right conclusions for the respective observer. That makes things a bit complicated, but it's the only way to get the right statements if we want to study physics in different frames of reference. What applies to masses, lengths and times, but does not apply to charges and speeds. The latter are preserved in the transition from a stationary to a moving frame of reference. While the moving observer does not perceive any changes due to speed in his local system with his local measurement system, the stationary observer observes some changes due to the energy changes in the moving reference frame. The absolute increase in the size of the Bohr radius r_v as a function of the velocity of the atom for the stationary observer *s* is:

$$r_v[s] = \gamma \cdot r_s[s] \tag{4.43}$$

where γ is given by equation (4.17).

Of course, this also has an effect on the de Broglie wavelength and thus on the energy levels in the atom, which is why, for the stationary observer, the emitted wavelengths shift towards longer wavelengths. It may come as a surprise that Planck's constant in the transition from a stationary to a moving frame of reference must change for the stationary observer because it contains the mass of the electron.

$$h[s] \cdot \gamma = h[v] \tag{4.44}$$

While we have been dealing with the inertial mass and have traced it from macroscopic viewpoint to the single atom. In doing so we discovered that mass is not invariant to physical transformations in terms of lengths and times. The question, what about the heavy mass and the gravitational force, still remains open. We will answer this question later, when we deal with the properties of the atomic nucleus. In addition to energy and mass matter also has a third player, the light with its propagation speed, which simultaneously represents an upper limit for the movement of masses, which can not be exceeded. This is what our first basic axiom of physics says about the preservation of matter in section 2.4. The four basic Axioms for Physics

4.5 The Puzzle about Light and its Speed

Everything that we have previously recognized about the physical properties of matter is based on our sensory abilities 'to feel' and 'to see'. The beautiful word 'to grasp something' comes from our sense of touch. This one is very short in its range. On the other hand, seeing is a long-range sense. Everything we can see and observe comes from the light, more specifically from the interaction of light with a phase transition of masses. Nevertheless, the light is still largely puzzling, because in the truest sense of the word, we can hardly grasp it. The fire, this fourth state of aggregation, warms our dwelling and illuminates it in the darkness. The thermal properties of the light have been known from time immemorial. On the other hand, it dawned on us that light had something to do with electricity since Edison began illuminating our houses in 1876 with the invention of the electric light bulb, after Werner von Siemens in 1866 of an industrial scale, succeeded in applying charge separation by means of a generator. As early as 1864, James Clerk Maxwell had formulated for this the equations for the theoretical basis of optics and electrical engineering. The electron as the cause of light as part of the electromagnetic radiation was first detected in 1897 by Joseph John Thomson as elementary particle and first determined by Robert Millican in 1907 as the smallest amount of free charge. Obviously, equation (3.11) has its justification as $E = h_e \cdot v$, since h_e is the electron's action quantum.

Light, like any other electromagnetic radiation energy, is generated by the action of the movement of an electrical charge.

But that started the fight. Is light a particle or a wave? The theory known as **corpuscular theory** is a physical theory attributed above all to Isaac Newton according to which the light consists of the smallest particles or corpuscles. The corpuscular theory was replaced in the 19th century by the wave theory of light. However, since the photon theory of Albert Einstein (1905), light has partly been attributed to particle properties. Einstein received in 1921 the Nobel Prize *for his services to Theoretical physics, especially for the discovery of the law of the photoelectric effect*. His work on the photoelectric effect containing the light quantum hypothesis, which already presented Newton and is today regarded as a pioneer work of quantum theory. Perhaps Einstein would have been forgotten if Max Planck, who was the founder of quantum mechanics and at that time Professor of Theoretical Physics and a member of the Prussian Academy of Science, had not noticed the young Einstein. Already in March 1906 Planck gave a lecture on Relativity in front of the Physical Society in Berlin and was in contact with Einstein, who was still living in Bern at the time. Since the introduction of quantum mechanics, they speak of the illogical dual character of light.

A central question for the character of light is also: **Does light need a propagation medium and in which direction does light spread?** If light is a wave then there must be a transmission medium and it will spread in all directions. If light would consist of particles, you could collect the particles and store light. Only in favor of the Theory of Relativity they have negated the transmission medium, but it has not been able proving the possibility of light storaging. In addition, particles have mass and we have seen that the mass appears to increase with velocity and it seems to become huge when approaching the speed of light because the equation does not take into account the inhibitory effect of the force field of this propagation medium. Consequently, light can not consist of mass particles. Although equation (3.11) states that the energy is transmitted through quanta of action, similar to hammer blows to a wall, it does not say that mass is transmitted. Only forces in the form of pulses are transmitted. There must be a medium that can transmit these powers, as forces are bound on mass. However, we can not grasp this medium with our senses because we have neither electrical nor magnetic senses. But we can measure magnetic fields, and where there are magnetic fields, there must also be vertical to it electrical fluxes. On the other hand, using laser resonators, you can force the light to propagate only in a certain direction. Does this property alone justify looking at light as a particle?

The energy flow is bound to a carrier within the matter. Light is a flow of energy and is bound to mass as a carrier. Only here does not the visible mass move, but there is an invisible power coupling around the mass, which guarantees the mass cohesion. This is the carrier of the light and the transfer rate depends on Kohlrausch and Weber [4.16] on the electromagnetic properties of the carrier.

The discovery of Kohlrausch and Weber leads to the spectacular relationship:

$$h \cdot \nu = m \cdot c^2 = m/\varepsilon_0 \cdot \mu_0 \qquad (4.45)$$

The first part of the equation is also called the Planck-Einstein relation. ε_0 is the dielectric constant, which indicates the permeability of the vacuum for electric fields, and μ_0 is the permeability of the vacuum. The permeability indicates the transmittance of the vacuum for magnetic fields, where the term vacuum is not clearly defined. Vacuum is supposed to be the absence, the emptiness of matter [4.17]. If that were true, equation (4.43) on the right side would yield zero. There would be no energy in the vacuum. Without energy, there would be no forces in the universe and the beautiful order would be gone. However, we have found that there is no absence of matter because matter is what exists outside our consciousness. Matter occupies a volume, no matter how small its density is. With no machine, we have ever been able to create a vacuum that contains fewer atoms than that what exists 500 km above our heads. Consequently, there must be more energy there than we are able to put into machine power. Even one atom per cubic meter is still matter, mass and kinetic energy. The term emptiness goes back to our sense of sight, which sees nothing, even thought forces can work. But the light penetrates through this optical 'void' where it is invisible. Only when it encounters a phase boundary does its effect as a change in velocity become visible as optical refraction or reflection. It acts like a tsunami that hits land. There the speed of the tsunami wave decreases and the consequence is the towering up of the devastating water wall. The dielectric constant and the permeability are indeed material constants, which influence the propagation speed of the light. If you compare these material constants with all other substances, then hydrogen comes closest to what corresponds to the spectroscopic findings about the frequency of hydrogen by universe. The light is described in physics with the laws of optics that we can rely on, as we build the devices that magnify our sense of sight many orders of magnitude. At the same time we go right into the theory of electromagnetism. The beam flux **S** of the light is thus determined by the electric field **E** and magnetic field **B** as the magnetic flux density and is

$$\boldsymbol{S} = \boldsymbol{E} \times \boldsymbol{B} \tag{4.46}$$

This equation looks pretty straightforward at first glance. It consists of three vectors connected by a cross product and **S** is perpendicular to the plane forming **E** and **B**. So does the light have a direction? It is the direction from its source to the observer. The observer can be anywhere. So equation (4.44) does not hold for one vector **S** but for a set of vectors and **E** and **B** are also sets of vectors. If we have sets of similar vectors, we call this a *vector field*, and if these vectors are functions of variables we describe these fields with methods of **differential geometry**, a consideration on a small scale. To understand force fields, we look for places equal field strength and connect them with lines, as well as one on a map the terrain profile with contour lines showing and examine from there the surroundings. Friedrich Gauss has found that the electric field lines diverge in presence of a charge, while the magnetic field lines do not diverge. The field of magnetic flux density **B** is free of sources; there are no magnetic monopoles. From this there is the partial vector operator div for divergence. This fact can then be expressed as follows: div $\boldsymbol{E} = \rho/\varepsilon_0$ with ρ as the charge density and *div* **B** = 0. Here, the relationship between the magnetic field strength **H** and the magnetic flux density **B** is:

$$\boldsymbol{B}/\mu_0 = \boldsymbol{H} + \boldsymbol{M} \tag{4.47}$$

M is the magnetization. The law of induction says: The changes in the magnetic field lead to an electric vortex field. The vortex field is described by another partial operator with the short cut *'curl'*. *curl* $\boldsymbol{E} = \partial \boldsymbol{B} / \partial t$ and the change in the electric field leads to a magnetic vortex field. *curl* $\boldsymbol{H} = \partial \boldsymbol{D} / \partial t$, where instead of **E** is the electric flux density **D**. The connection between **E** and **D** is given by:

$$\boldsymbol{E} = (\boldsymbol{D} - \boldsymbol{P}) / \varepsilon_0 \qquad (4.48)$$

P stands for the polarization. From Friedrich Gauss comes div (***D*** - ***P***) = ρ. The displacement current, which otherwise has to be taken into account in the law of induction, plays no role in the transmission of light. From these formulas, the relationship between the electromagnetic character of light and the general energy formula describing matter can be seen.

We introduced the concept of differential geometry above. Now let us consider the operation of the operators div and curl on the vectors E and H, and D and B, respectively. These four operator equations are known by the name Maxwell's equations.

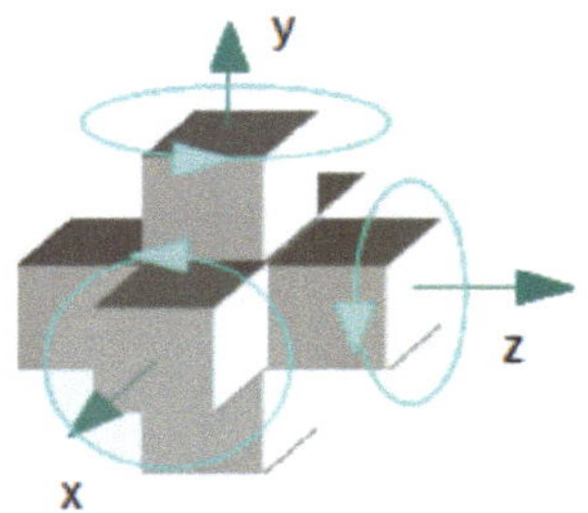

Figure 4.3: von Neumann neighborhood in space

The differential geometry looks at the world on a small scale from a spatial cell and it is obvious then that it is interested in its neighborhood. We already have got to know such an approach under section 3.4.1 Causality and Coincidence. where we introduced the Conway game. A physical volume, as opposed to a mathematical space, is always characterized by neighboring atoms with their force field. Electromagnetic fields are also material because they are carriers of energy. Without loss of generality, we can therefore divide the space into neighboring cells and examine the effect of a cell with certain properties on its neighborhood by imagining that in each space cell, an atom provides the necessary field. We have here, for the sake of clarity, based on a very simple neighborhood relationship. the '*von Neumann neighborhood*'. John von Neumann was one of the pioneers of computer engineering and an excellent mathematician.

To make the operation of the Maxwell's equations visible, we elect the principle of Conway game and translated the operator equations of Maxwell in to game rules.

- The first rule of the game is the divergence. Electric field lines diverge under the presence of electrical charge. Magnetic field lines do not do that. Divergence is a rule that describes the formation of a field. The div-operator identifies this rule. It describes a change in a field, here in our case a shift along the coordinate axes..
- The second rule is the rotation. The rotation is a rule that generates a new field vector perpendicular to the summed vectors by summing field vectors of a field in the neighborhood of a cell in the counterclockwise direction. We call the operator marking this play rule, *curl.*

The Maxwell equations are then:

$$\begin{aligned} div\vec{E} &= \rho/\epsilon_0 & rot\,\vec{E} &= -\frac{\partial\vec{B}}{\partial t} \\ div\vec{B} &= 0 & rot\vec{B} &= a\left(\frac{\partial\vec{E}}{\partial t}\right) \end{aligned} \tag{4.49}$$

Because only the effect is transported, eliminates the displacement current at *curl*. The time is realized in our game by a step on z-axis.

When a quantum of light is generated, we know that an electron jumps from an energetically higher orbit to an energetically lower orbit. However, the electron has to deliver as much energy in a short time as the energy difference of the two paths is. It can do this only by swinging back and forth between the two orbits, creating an electromagnetic wave train whose energy content corresponds exactly to the energy difference of the two electron orbits. When an electron moves outside the electron orbit around the atomic nucleus, this is noticeable by a tiny flow of current. The current flow generates a Lorentz force perpendicular to the current flow, which shifts the electric field lines. We describe this with the game rule *div* at the position $z = 0$. The rule of the game works in such a way that due to the Lorentz force all available vectors on the z-axis are shifted to the right by steps one unit π/12. Then, in our game at z = 0, a new vector is created, whose amount goes in steps through a sequence of values 0,0.23, 0.45, 0.65, 0.80, 0.92, 1.00 to π/4 of move to turn.

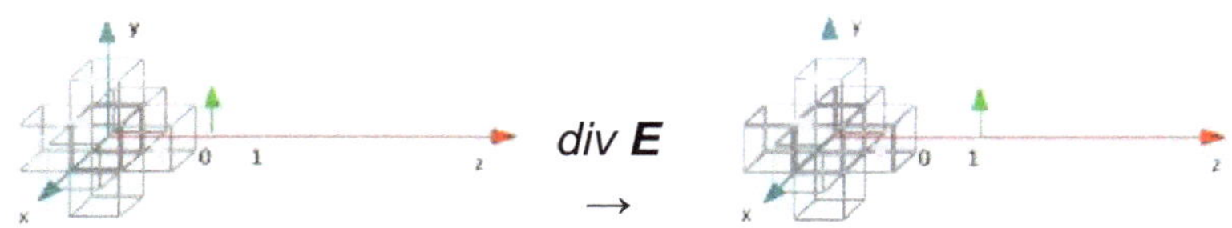

Figure 4.4: Effect of the game rule div ***E***
Here we have chosen the preferred direction +z. Because div ***E*** *is not a vector, every direction is equal. The propagation is spherically symmetric according to the radial E field. That should always be kept in mind. Since light is always perceived by all observers as if it came directly from the source as a ray, only one direction is of interest to each observer.*

Now we have to deal with the effect of the game rule *curl*. For this we look at the ***E***-vectors in the neighboring cells of cell (0,0,0) in the center and add off the vectors counterclockwise. The neighboring cells (0,1,0) and (0, -1,0) never contain field vectors, because the wave should propagate only in the z-direction and the excitation center lies in the cell (0,0,0). We then get 6 different cases of how the ***E***-vectors can be directed around the cell (0,0,0). There is a rule of thumb for the addition. When the fingers of the right hand point in the direction of rotation, then the thumb indicates the direction of the new vector. Figure 4.5 shows the orientation of the ***B***-vector for the different cases. In the case of a homogeneous field, we get *curl* ***E*** = 0, which shows the two pictures in the middle row in Figure 4.5.. In the other cases, the magnitude of the vector is different from zero, respectively.

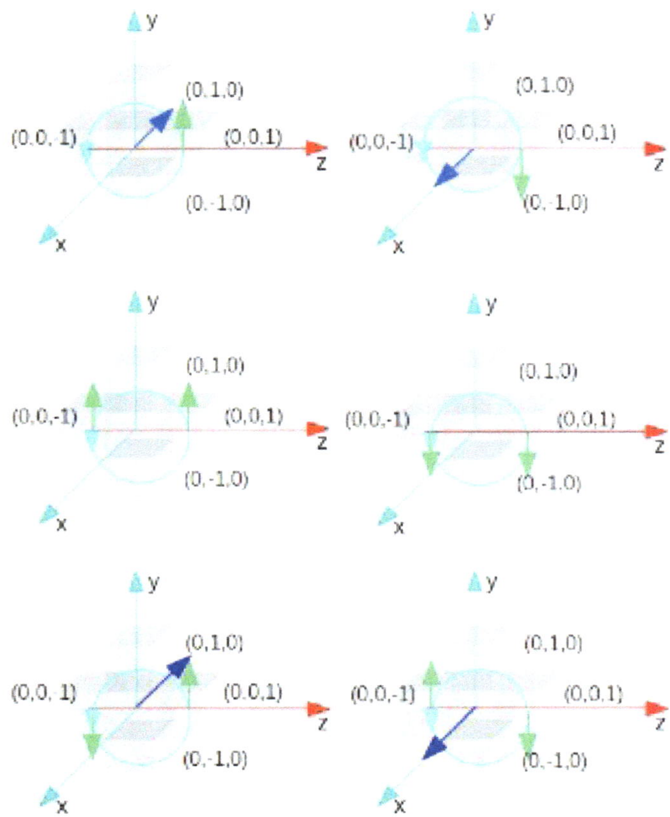

Figure 4.5: Effect of the game rule curl ***E*** *is indicated by the blue arrow*

The blue arrow in Figure 4.5 indicates the result of the rotation. The direction of the result vector is the magnetic field strength. This vector is perpendicular to the electric field strength. However, our rule is *curl* ***E*** = -∂ *B*/∂t, (**The operator Δ stands here for** *∂* **/∂t**), the negative sign which is why the orientation of the *B* vector is mirrored to the x- direction.

The series of Δ***E*** amounts change sinusoidally in π/12 increments. Then we get *curl* ***E*** = - Δ***B*** = -((0, -1,0) -(0,1,0)) for the half-wave following amounts:

π/12	0	1	2	3	4	5	6	7	8	9	10	11	12
ΔE	0,00	0,26	0,50	0,71	0,87	0,97	1,00	0,97	0,87	0,71	0,50	0,26	0,00
−ΔB	-0,47	-0,25	-0,23	-0,18	-0,13	-0,07	0,00	0,07	0,13	0,18	0,23	0,25	0,47

The magnetic field has no source like the electric field, so *div* ***B*** = 0 and we do not get a game rule. Now we have to apply -Δ***B*** the 3rd game rule *curl* ***B*** = *a*·Δ***E*** = *a*·((0, -1,0) – (0,1,0)) to the above line and we get the following values!

π12	0	1	2	3	4	5	6	7	8	9	10	11	12
ΔE	0,00	0,07	0,13	0,19	0,23	0,26	0,27	0,26	0,23	0,19	0,13	0,07	0,00
aΔE	0,00	0,26	0,50	0,71	0,87	0,97	1,00	0,97	0,87	0,71	0,50	0,26	0,00

The constant *a* in our model depends on the size of the cells. We determine *a* such that the amplitude of Δ***E*** becomes at most 1. Then we get *a* = 3,732. This is necessary so that the amplitude of the wave remains constant. In reality, *a* is the product of the electrical field constant *ε* for the electrical properties and the magnetic field constant *μ*, which describe the electromagnetic properties of the medium.

$$a = \epsilon \cdot \mu = \frac{1}{c^2} \tag{4.50}$$

You can reduce the cell size arbitrarily, for example, up to 5×10^{-7} m in the order of the wavelength of visible light. But that does not change the principle and does not mean that our environment is discrete. In any case, it remains only a normalization factor. Now we need the orientation of the vectors E from the rule curl ***B*** = *a*·Δ ***E***

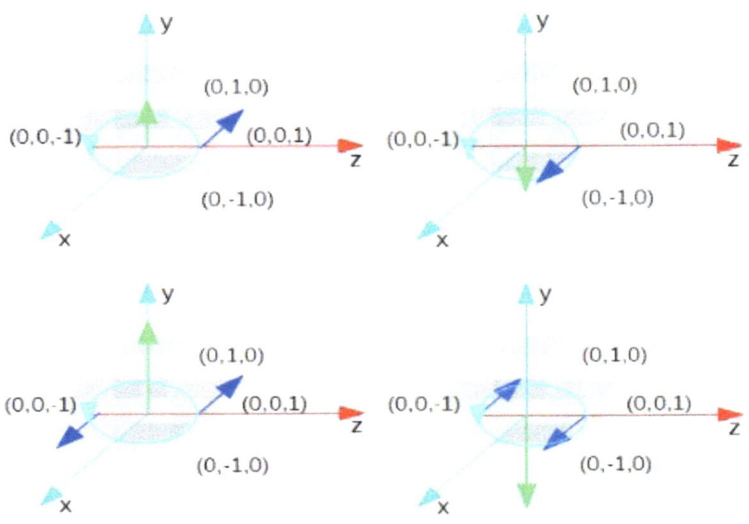

*Figure 4.6: Effect of the game rule curl **B** is indicated by the green vector*

We have already discussed the transmission of an impulse in section 3.4.5 Planck's Constant and the Domino Day.. From the Maxwell equations and the Weber-Kohlrausch relation follows the relation of the speed of light with the electromagnetic force field of a for this spectral region permeable mass state. At the moment when the wave encounters an obstacle due to a phase change (suppression of permeability), a 'tsunami effect' is created by congestion of the energy flow and this energy discharges in the form of an effect quantum at the location of the receiver. The residual energy is reflected.

Now let's experience rules 1 through 3 in context:

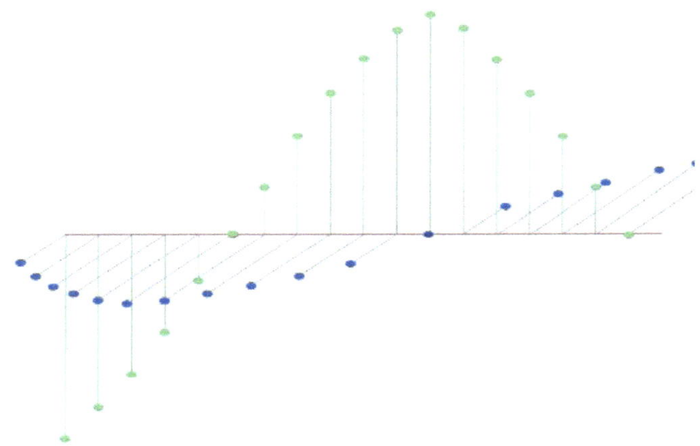

Figure 4.7: Model of an electromagnetic wave according to the above rules of the game (animated)

The secret of the propagation of the wave train lies in the coupling the three rules above! The effect of this propagation occurs at the phase boundary.

How grotesque it seems when someone comes along and wants to symmetrize these equations?

> *»It is known that Maxwell's electrodynamics—as usually understood at the present time—when applied to moving bodies, leads to asymmetries which don't appear to be inherent in the phenomena.«*

like Albert Einstein's momentous essay *On the Electrodynamics of Moving Bodies* begins [4.12]. In this essay he uses the Lorentz transformation, a kind of projective transformation, which is a mapping to eliminate the game rule *div E*. This is equivalent to putting the observer on the light beam. This violates the principle that no mass can ever

reach the speed of light. He also demanded that the speed of light must be constant. This requirement contradicts the laws of optics, which his disciples regard as a proven fact. And what the hell is the symmetrization that destroys the dynamic function of Maxwell's equations? Genius and madness have been two sides of the same coin since Plato. Here madness was held by society for genius, and so fate in physics took its course, since his model was to make school. Einstein as a sorcerer's apprentice had started a process that could not be stopped anymore. But still worse, he continued his work with his General Theory of Relativity. We will discuss this in the chapter 6 The Macrocosm.
But before that we have to ask: What does it mean if the case $div\, \mathbf{E} = 0$ actually occurs with no transformation. Then we have two stable interlaced vertebrae. This results in a permanent magnet. How small nature can make permanent magnets? We explore the question in the following chapter 5 The Microcosm.

It is amazing how little physical understanding prevails in the world's colleges that today this pseudo scientific theory called special relativity theory is still taught though it is only a special view. Even in Jena, where Otto Schott, Ernst Abbe and Carl Zeiss laid the foundation for the optical industry in the late 19th century, through whose work we are today in a position to explore the depths of the cosmos, University established a chair of relativity. All our information from the cosmos we receive by means of light as an optical information carrier. No single light source from these spheres has given us primarily information about any forces.

4.6 Interaction of Light with the Mass

If the equation $E = h_e \cdot v$ is accepted as correct, you can not interpret v as the frequency of a wave in the known sense as a motion of coupled particles, but must understand it as packet of few oscillations between two coupled force fields of which one is electric and the other is magnetic. The nature of these two fields is described by Maxwell's equations. When we think of waves, we usually have the image of a surface wave in mind. A better model is a tsunami wave. This wave is not perceived on the high seas at all. Only off the coast does it release its energy, released as a result of a sub-sea vibration. This is an effect that occurs at the boundary of two phases of mass.

When light hits mass,.it can interact with the atomic shell in two ways.

- The first type is **Thomson scattering**, a scattering at the electrons, without an impulse being transmitted to them. The light is scattered in all directions, causing the sky to turn blue.
- The second type is the **Compton effect.** This has been experimentally proven only on high-energy radiation. Here, a light pulse is transmitted to the atomic shell and re-emitted. This is accompanied by an energy loss of the incident radiation. An electron is accelerated for a short time as a result of a pulse transmission. The acceleration of an electric charge induces electromagnetic radiation according to Maxwell. P. Marmet has shown that this phenomenon has a very large cross section in the forward direction and that the energy lost by bremsstrahlung causes a slight redshift of the incident light. He also showed that this radiation effect is indistinguishable from the Doppler effect of light already discussed in section 3.6.2 [4.18].

When light is absorbed and re-emitted in a mass flow, the short moment between absorption and re-emission should have an influence on the speed of light. As early as 1810, François Arago experimentally examined the possibility of the influence of the movement of a prism on the refraction of light, which should lead to a changed aberration angle[3]). According to the then accepted corpuscular theory of light, he expected a change in this angle in a measurable magnitude, as the veloc-

3 The aberration angle depends on the current position of the earth on its orbit around the sun. It alternates its sign twice a year and oscillates between a_{max} and $(-a_{max})$. Because the velocity of the earth v is perpendicular to the beam direction, you obtain for $\tan(a_{max}) = v / c$

ity of the incident starlight adds up to the earth's velocity on its way around the sun. However, there were no measurable fluctuations in the deflection angle as a result of refraction during the course of the year. Arago explained this result by arguing that starlight is a mixture of different velocities, whereas the human eye can perceive only one of them [4.19].

Fresnel (1818) explained the result of Arago with the assumption that the speed of light within light-transmissible bodies is modified by partial entrainment of the light-conducting medium, the ether. This entrainment would be due to the fact that in the bodies the ether is compressed and therefore a little denser, whereby exactly this excess of ether density - with the exception of the range of the normal density - is carried along by the bodies. The *Fresnel dragging* coefficient Φ (where *v* is the velocity of the medium and *n* is the refractive index) is given by:

$$\Phi = \left(1-\frac{1}{n}\right)v \qquad (4.51)$$

The refractive index n is the ratio between the speed of light in vacuum and the speed in the stationary medium. This leads to the conclusion that the velocity reduction in the medium is due to the interaction of the light with the atomic shell as it travels between the atoms at the vacuum velocity. Only during this brief moment between absorption and re-emission can occur any influence on the speed of light by flowing atoms.

$$\Phi = \left(1-\frac{{c_M}^2}{c^2}\right)v \qquad (4.52)$$

An accurate confirmation of the coefficient of contribution was made possible by the Fizeau experiment by Hippolyte Fizeau (1851) [4.20]. He used an interferometer arrangement in which the speed of light in the water was measured. The construction of Fizeau's experimental setup consists of two light paths within water-filled tubes. The water flows through the tubes, in such a way that the first ray of light passes in the direction of the flowing water at the speed *v*, the second ray of light runs against the water flowing at the velocity *-v*. When merging the beams in the interferometer results in a *v* dependent path difference, which can be seen by a shift of the interference pattern. The speed was measured at a path difference for odd multiples of λ/2, because here the minimum of the interference fringes occur. Classically you would assume that the speed of light in the medium is:

$$c_M = \frac{c}{n} + v \qquad c_M = \frac{c}{n} - v \tag{4.53}$$

in flow direction and against

However, speeds were measured which were smaller by the factor (1-$1/n^2$) than the speed calculated for the path difference λ/2. The result thus says that the speed increase or decrease of the speed of light in the moving medium is not additive, but that:

$$c_M = \frac{c}{n} + v \cdot (1 - 1/n^2) \qquad c_M = \frac{c}{n} - v \cdot (1 - 1/n^2) \tag{4.54}$$

in flow direction and against

This means that for $n \rightarrow 1$ is $c_M = c$, which was interpreted as a dormant ether. There was a quarrel between the representatives of a stationary and a moving ether. If the ether rests, one would feel the ether wind like in an open car. Michelson and Morley conducted a first experiment in 1896 for this purpose. The result was at the detection limit of 4-8 km/s and therefore was declared to zero. The speed of the earth around the sun was expected to be just under 30 km/s, which according to equation (4.50) can not be. In particular, Dayton Miller was a representative of the ether theory of that time, and a particularly persevering and thorough experimenter who believed that the Fresnel dragging coefficient increased in altitude, which is why he relocated his investigations from laboratory basement to Mount Wilson. Around 1933, Miller discovered that the earth drifts at a speed of 208 km/s towards the vertex in the southern hemisphere of the sky, towards the swordfish in the middle of the Great Magellanic Cloud and 7° from the southern pole of the ecliptic [4.21]. According to recent measurements, the drift speed of the solar system in the milky way is about 225 km/h.

To evaluate the dispute, the following table shows the Fresnel's coefficient of carry-over (4.48) for a terrestrial experimental set-up at some typical cosmic velocities.

Fresnel Dragging coefficient by	Speed of Earth around Sun	Speed of Sun in Galaxy	Galaxy
Air on mare level	17,4m/s	174m/s	583m/s
8000m over mare level	6,3m/s	63m/s	220m/s

Table 4.1: Fresnel dragging coefficient of entrainment for characteristic cosmic velocities

Assuming a Fresnel dragging coefficient, the calculated flow velocity of the ether remains more than an order of magnitude below the values given by Michelson and Morley. In addition, according to the entrainment theory, the speed change would have to decrease with altitude. This shows that the mechanism of light discovered by Miller has nothing to do with the previously known properties of light. Robert Shankland (a student of Miller) was commissioned by Einstein to submit Miller's data to statistical analysis. According to James DeMeo, this would not have been done very carefully and only confirmed Einstein's prejudice as to temperature influences. Today, all the textbooks contain only the experiment of 1887 conducted by Michelson and Morley. The assumption of a resting ether seems to contradict the result of the Michelson-Morley experiment, in which the evidence of Earth's motion relative to this ether failed. Based on his analysis, Paul Marmet concludes that the Michelson-Morley equations of the last century did not consider that the reflection angle of light on a moving mirror is a function of the speed of the mirror. This applies to both beam directions. Taking these previously overlooked phenomena into account, the back-and-forth reflections cancel each other out, and so Marmet shows that an invalid result of the Michelson-Morley experiment would be the result of an *absolute* reference frame (called *aether*, or field) of classical physics [4.22]. Nevertheless, Miller has measured an anisotropic effect in the direction of solar system motion. For an explanation, see Chapter 7 Outlook in a Intergalactic World

Then Miller would have fitted better with his moving ether to the theory of relativity? Certainly not, because his results do not confirm Einstein's requirement for a constant speed of light independent of the movement of the ether! Albert Einstein, wrote in a letter to Edwin E. Slosson, July 8, 1925:

> *»My opinion of Miller's experiments is the following. ... If the positive result is confirmed, then the special theory of relativity and with it the general theory of relativity in its present form will be invalid. Experimentum summus judex. Only the equivalence of inertia and gravitation would remain, but they would have to lead to a significantly different theory. «*

This statement refers to the constancy of the speed of light, which is necessary for the Lorentz transformation and would not work in a moving ether.

The alleged failure of the Michelson-Morley experiment is regarded as a necessary requirement of the theory of relativity!

A report on the refutation of the Miller experiment gives the Hammar experiment [4.23]. It is an experiment in which a partial beam passes through a lead chamber. This was to show that the deadweight effect does not exist. However, according to the above table, it would have to be almost three orders of magnitude greater accuracy than Miller's experiment to exclude the entrainment, which is even more doubtful, since the effort for eliminating the disturbance would have to be correspondingly higher than Miller. All this shows the passion with which was discussing this Question

We can easily imagine the failure of the Michelson-Morley experiment on the ground, if we transfer it in our mind to a closed moving train and ask if we can measure the speed of the train by the fact that we are running in each case a distance in the direction of train travel, and take another course running perpendicular to the travel direction at a constant speed and comparing the covered distances.

However, if you send the two parts of a split light beam through a rotating apparatus so that one beam is reflected with the direction of rotation and the other against the direction of rotation, you get a path difference; since due to the rotation of the apparatus, the two circumferential paths are different lengths because of the superposition the speeds. Sagnac succeeded for the first time in 1913. Today the Sagnac effect is used technically in optical gyroscopes.

The difference between the two experiments is that in the Michelson experiment the light source don't move in the earth's field, with respect to the observer but not in the Sagnac experiment. Consequently, in the second case, a superposition of the speed of light with the speed of the light source is observed. This is also the case with the Doppler effect. Here one observes an extension of the wavelength in the case of a removing light source. A moving observer can not see these effects.

More recently, the Michelson-type experiments have revived: on February 11, 2016, the LIGO[4]) team [4.24] published the observation of alleged gravitational waves from a black hole. The LIGO experiment uses two observatories to record gravitational waves. Both are at a distance of 3000 km in the US. In the two observatories, laser beams pass in two Fabry-Pérot resonators at right angles to each other, which form a Michelson interferometer with an arm length of 4 km. Experimental evidence for the existence of a gravitational wave has been announced: Mirrors of 40 kg mass should have been displaced by 10^{-18} m in fractions of a second, measured with a Michelson interferometer. Scaling

4 LIGO = **L**aser **I**nterferometer **G**ravitational-Wave **O**bservatory

up these data by a factor of 1013 requires relative accuracy of a hair's breadth (10 microns) relatively to the distance to the nearest star (4 light-years). This would be an accuracy that would be better by a factor of 1 million than the accuracy of a Mössbauer spectrometer, which has so far reached a resolution of 10^{-15} m. The synchronization of both observatories should have separated the interference from the useful signal. Considering the evaluation of the measured data and its preparation, as Walter Orlov [4.25] has done, one concludes that the 'glitches' filtered out of the background noise must be tilting oscillations, as they already found in resonators in 1907 by Heinrich Barkhausen [4.26]. If such a glitch accidentally happens to occur almost simultaneously in both resonators, the coincidence is interpreted by the LIGO experimenters as a signal of a gravitational wave triggered by the fusion of two black holes. This is an unquestionably dubious interpretation, driven by the desire to show anything at all. Where is there a causal connection between the glitches in a background noise and any imaginary cosmic phenomena whose existence is like a theological argument not proven at all and only exist in mathematical fantasies?

4.7 The Entropy

What is entropy? This basic concept of physics is generally less well known and it causes the most difficulty in terms of understanding. We briefly introduced it in Equation (4.05). It stands at the interface between macroscopic and microscopic observation of physics. Its macroscopic side was discovered by Rudolf Clausius as early as 1865. He introduced the term and since he has been considered the discoverer of the 2nd theorem of Heat Theory, that states, that in a closed system heat from a colder body never can flow to a warmer body. Everyone knows, however, the principle of the heat pump. However, this is an open system that requires energy. We suspect that this Second Law of thermodynamics, with its limitation on closed systems, is incomplete.

At the beginning of the 18th century, the steam engine was invented, which converts a temperature difference into mechanical work. At that time, physicists tried to understand the principles that these machines obey. The researchers found out with irritation that only an about one third of the thermal energy could be converted into mechanical energy. The rest was somehow lost - without understanding the reason. Unlike the hitherto prevalent opinion, Clausius recognized that heat is not an immutable substance, but only a form of energy that can be transformed into the other known forms (kinetic energy, etc.). Robert Mayer postulated on the principle of energy preservation. However, the principle of energy preservation does not yet explain the common fact that energy conversion does not take place in every direction: why, for example, two differently warm bodies adjust their temperatures when in contact, but never heat goes over from the colder body to the warmer body. Even Carnot had made this fact clear, but first Clausius recognizes behind it an energy flow and not a phenomenon bound to a thermal substance. (See Figure 4.1)

The change in the amount of heat *dQ* related to the heat transfer temperature T in a thermodynamic process is thus a measure of the convertibility of heat into technical work and thus of the quality of the process, if we dissolve equation (4.05) for ds, and understand the energy flow as change in the amount of heat.

We get $dS = dQ / T$ in joules per degree.

Clausius later called this equivalent value of transformation *entropy* (from ancient Greek: entrepein = transform and tropé = change potential). The maximum usable, free energy in an isolated system is always smaller than the actual internal energy. Although the internal energy of the system is preserved in the conversion into useful work (1st principle), it is devalued, since any part is scattered or dissipated into the system environment. Thus, the second law of thermodynamics can also be formulated as follows: An energy conversion never goes by itself from a low-quality state to a high-quality state; the entropy is always increasing. In the heat-power process, heat from the outside (firing) must be used to "refine" the process medium, water, by producing water vapor under high pressure and temperature, before it will work in the cylinders of the steam-engine or in the turbine, to generate electricity. The energy of the leaving steam is worthless and must be released through the radiator into the environment. Even under ideal conditions, the production of scattered energy, such as waste heat, would be unavoidable.

The microscopic view of entropy was introduced by Ludwig Boltzmann in the second half of the 19th century when it became clear that heat has something to do with the movement of atoms and molecules. He focused on the microscopic behavior of a fluid, a gas or a liquid. He understood the disorderly movement of atoms or molecules as heat, which was crucial for his definition.

Potential energy → Kinetic energy + Heat

Heat is the chaotic motion of molecules and atoms. In a closed system with fixed volume and constant particle number, Boltzmann stated, the entropy is proportional to the logarithm of the number of micro-states in the system. Under micro-states he understood all the possibilities of how the molecules or atoms of the caged fluid can arrange. His equation thus defines entropy as a measure of the "freedom of arrangement" of molecules and atoms: as the number of ingestible micro-states increases, entropy increases. The entropy of a macro-state is calculated by the probabilities of the micro-states. The proportionality factor k is referred to today as the Boltzmann constant.

Using Boltzmann's definition, this side of the term can be understood - but entropy yet has another aspect. Often, entropy is referred to as a measure of the disorder of a system, but physics doesn't know a measure of disorder, which is why it says better:

The entropy is a measure of the ignorance of the state of each individual particle, which measures the information in order to be able to conclude from a known macro-state on the micro-state of the system.

In 1948, Claude E. Shannon tie up, in his fundamental work *A Mathematical Theory of Communication* [4.27], coined the modern information theory by defining the structure of the energy flow as an information-transferring. *Information* only becomes tangible when it encounters a phase boundary in the form of a detector. Susskind once spoke of entropy as hidden information [4.28]. Shannon made information measurable by removing it from the sphere of the recipient's personal interest. The basic idea of information theory is to regard the generation of information in a news source as a random process. In fact, information only appears in conjunction with an energetic or material carrier. Similar to the distinction between kinetic and potential energy, we distinguish between animated and dead information. The former is bound to an energetic carrier and the latter to a massive carrier. The informal entropy is measured in bits. These few remarks should suffice here. The fundamental meaning of entropy will only become clear in the course of the following chapters when we deal with energy flows through open systems. The information content of the energy flow is the subject of cybernetics and information theory.

Here it becomes clear that time has an informal character because it is bound to a clocked flow of energy in a physical system. In terms of the concept of time, however, we are not interested in the clocked flow of energy, but in the information on how often the cycle has been run in comparison with another movement. A single pendulum motion of a clock then corresponds to exactly one bit of information. However, for the pendulum to move constantly, it needs an external impulse, which is guaranteed by an external energy flow. So time is being de-mystified. We return to the fundamental importance of entropy in section 6.6.

4.8 What is Form?

The term of substantial or even structural form is not a term of physics because it is not a quantitatively measurable property. This is like in the special price hardware store in my town. There screws are sold in mass and not in pieces of shape. I can pack my assortment according to my need for quantity of screws, nuts and washers together in a bag. At the checkout the bag is weighed and then the price is determined. In physics, as long as one is in the macroscopic range, one proceeds similarly. One abstracts from the form and the number and is only interested in the mass of objects. The difference to the amount is that the mass is not counted, but compared to a reference mass, because the counting of large numbers is practically impossible to do.

The term 'substance' appears for the first time in chemistry. Substance and body shape are not basic properties of physics, but structural properties of matter, and thus reserved for the other sciences after splitting of philosophy in different scientific disciplines. However, according to the above, we can interpret the structural form as dead information and bring it close to entropy. Forms or shapes are visible at phase transitions of masses due to different reflection and diffraction of the light.

Nevertheless, there a group of non-specialist critics of modern physics around Bill Gaede [4.29] insists, as well as Raphael Haumann on the fact that form would be a fundamental concept of physics. They distinguish objects with forms as well as existences with their form and localization and confront them with concepts and build thereon their criticism of physics. Compared with the example of screw sales, this would mean that my bag would have to be opened at the cash register and the price of each individual piece would must be determined from its picking compartment. This would cause long queues at the cash register and sales would decline.

On top of that, the argumentation of these critics is that they do not view the mass as a representative of an object or a collection of objects, but rather only as a concept and therefore they deny mass a real existence. This is indeed ridiculous, because anyone who checks his own weight, in fact is destining his mass and annoying when he has gained weight on more fat , burdening his real existence. Such a reasoning is rightly rejected as pseudo-scientific, because it does not help, if one industriously collect citations worthy of criticism from the world of physics and this criticism then goes past the crux of the matter.

Shapes can be examined mathematically using the fractal geometry. The first one to study its nature intensively was the mathematician Bernoit Mandelbrot [4.30]. For him, shapes are a middle ground between a surface and a body. He speaks of a broken dimension between 2 and 3. So the shape is a mathematical term. But the people around Bill Gaede directly want to polemicize against the use of mathematics in physics by means of the concept of shape.

Physics can measure objects in terms of density, mass or temperature, so physics can distinguish objects only on that. Therefore, wanting to fix a critique of physics on the distinction between real objects and mental concepts on the basis of shape is not a productive critique. It is destructive because, in its logical consequence, it does not encounter the excesses of theoretical physics, which are not causally related to mathematics, but in their unqualified application to physical processes, as we have already seen and we will see in the next chapters. With a general enmity against mathematics, the basis of physics is hit. However, since Gaedeism essentially intends to hit the theory of relativity, in the end it is inconsistent and does not penetrate to the essence of what is currently getting out of hand in physics. We want to look at that in the next chapters and to contrast a different concept to relativistic physics .

A mathematical aspect of shape is geometry. While Euclidean geometry is the geometry of the real space and has always been rigorously axiomatic, an axiom always made headaches because it did not seem to be elementary. It was the phrase of non-intersecting parallel lines, or of intersecting at least in infinity. The sentence can also be replaced by the statement that the sum of angles in the right-angled triangle of a plane is 180 degrees. (In thinking, only extend the two longer thighs, then the acute angle becomes smaller and smaller and the two long thighs approach the parallels.) Can triangles be obtained with different angle sums?

We take an orange and cut a triangle out of its surface. If we apply tangents to the cut edges and measure their angles with each other, we get an angle total greater than 180 degrees. Why? Now the surface of the orange is a curved surface. On curved surfaces Euclid's parallel axiom is no longer valid. We call geometries that do not refer to space but to surfaces, non-Euclidean geometries. All of these surfaces have a curvature. The geometry of the Earth's surface is a non-Euclidean geometry. This must be taken into account when projecting large-scale maps onto a single plane. It is to be carefully distinguished between surface and space. A surface can be described with a function but a mathematical space not.

One physical aspect of shaping is the transition between two phases of one or more substances. One of the best known examples is the crystal formation from a liquid. Another is vortex formation by deceleration on a denser medium in a thinner medium. Even in our Conway game of life, spontaneously stable or oscillating figures appear as long as they do not hit the edge of the board. But the chaotic behavior is more common. Decisive for structure formation are probably coupled rules or partial differential equations that can be modeled with cellular automata whose behavior is determined by their environment and their internal state. The typical swarm behavior takes shape. Crystallization from saline solution or the formation of snowflakes evidently results in the removal of heat, which corresponds to entropic withdrawal in an open system. Even if the simulation of special processes is successful, we can not say that these processes are already understood in theory. Complexity, in particular, poses the decisive difficulties for linear causal thinking. But with the help of simulations on powerful computers, the solution to such problems is moving over closer and closer.

Now that we have discussed the terminology of basic physical concepts which are the objects of physics, let us turn to the physics of the microcosm to uncover the causes of forces and movements.

5 The Microcosm

> *»The majority of physicists simply refrain from looking into the meaning of their quantum calculations. Nor can they be disturbed by the consequences of the basic quantum mechanical experiments. Science does not have to do with truth or meaning for them, but with predictions and control: it is an instrument. Here will from the scientific arrogance of the 19th century, the cynicism of the 20th century. We have power, and that is enough for us; we can completely renounce real knowledge...« -* Arthur Zajonc

In the previous chapter we dealt with the terminology of matter, and now we want to penetrate into the depths and concern our-self with the smallest particles of the matter and their cohesion. The microcosm is usually described with quantum mechanics. We have under 3.4.2 Determinism versus Statistics. found that behind quantum mechanics lies a statistical approach. But physicists do not like statistics, which is why the Copenhagen interpretation is preferred by the most. »*According to this interpretation, the probability character of quantum theoretical predictions is not an expression of the imperfection of the theory, but of the fundamentally undetermined character of quantum physical natural processes,*« reads WIKIPEDIA as an explanation of the Copenhagen interpretation. Obviously, there are doubts about the theory after all. But with the Copenhagen interpretation, conditions are turned upside down and physical determinism is linked to in-determinism. How nonsensical that is, proved by a simple combination of statistics and determinism: "The brook is on average 50 cm deep and still the cow is drowned in it." We first have a statistical statement. This statement does not mention that the brook can also have deep spots and in other places it is very shallow. The 50 cm was understood as a middle value. It is not the vagueness of the brook depth, but simply the ignorance of the local conditions which led to the accident. At the ford the cow could have safely crossed the stream. We can not distinguish our ignorance from real existing in-determinism.

Conway's game of life reveals this dilemma very impressively. It has three very clear rules and yet very different results come out. The same is true of the three-body problem of cosmology. There is a whole problem class, which is summarized under the term chaos, whose rules are incomprehensible to us, because we can only think in causal rows. Here, however, causal relationships are often networked. Chance or accident arises where two causal rules meet. This is also the dilemma in the search for a combination of statistical approach and deterministic approach, like it is the aim of quantum gravity theory. As Brian Greene wrote it in his book *The Substance of the Universe* [5.01]:

> *»We only can hope to understand the origin of the Universe - one of the most important questions in science at all - if we succeed, the Conflict between general relativity and quantum mechanics to solve. We must overcome the contrast between the laws of the great and the laws of the small and unite them in a single, harmonious theory.«*

This hope will not be fulfilled. Physical theories are intended to provide tools for understanding nature, and to that extent they must be adapted to the object to be worked on. For an engineer, it is incomprehensible why we should work large and small things with the same tool. But perhaps this aim is only an expression of a general aimlessness of theoretical physics of the modern age, because the engineering sciences today set the aims and realize them.

When we divide mass, we always get again mass. This division can be continued in areas of 10^{-11} m. Then we reach the sphere of the individual atoms. With this we go over to count-ability and from the mass we get a set of protons (and neutrons). The electron microscope allows only magnifications of 1: 106, so the atoms remain well below the visibility limit. In addition, each mass has a certain temperature, which is to be understood as the vibration of the atoms, or their movement. Nevertheless, there are physical methods for determining the size of atoms, as H. Niedderer & J. Petri explain [5.02]. At the same time, they note that a definite atomic radius can not be specified because of the asymptotic behavior of the wave function of quantum mechanics. If you divide further, you get a negatively charged atomic shell and a positively charged nucleus. To determine the nuclear radius, we use high-energy electron beams, directs them to the atomic nucleus and observe the scattering behavior. We get an equation [5.03] for the approximation

$$r_{Kern}=r_0\cdot\sqrt[3]{M} \tag{5.01}$$

r_0 is the proton radius. Recent measurements of the proton radius gave a value of 0.84184×10^{-15} m [5.04]. The mass number M is the nuclear mass in units u, which corresponds to the mass unit of a proton [5.05]. The atomic mass unit u is defined as 1/12 of the mass of a carbon atom of the carbon isotope ^{12}C, i.e. the nucleus with M = 12 including the 6 shell electrons of the atom. So you can calculate with u equal 1 The electron has a three times bigger radius of 2.81×10^{-15} m.

The electron is thus more than three times the size of the proton, but has 1896 times less mass and both are the only stable particles.

This means that all other discovered short-lived particles are obviously to be understood as excited states of these two ground states, which is why, in these considerations, we will focus exclusively on these two types of particles.

While electrons were recognized as elemental, the positive antagonists had been a wide field of investigation and closed for a long time to physics. Only mass spectrometry brought light into the world of inorganic chemistry. The proton crystallized out as the building block of the atomic nucleus, which provided the mass in the core. Nuclear fission produced another particle with no electric charge, the neutron. However, this particle proved to be unstable. After a mean half-life of 15 minutes, it turned to a hydrogen atom. (After 15 minutes, there were only half of the initially available neutrons and after another 15 minutes, there were only a quarter and so on. From the neutral particle two particles of opposite charge were created.) In the aftermath, the high-energy particle physics reported the discovery of ever new particles. But none reached only nearly the lifetime of a neutron. Is it possible to speak of autonomous particles with lifetimes of the order of 10^{-6} s and shorter, or are they only excited states of known particles? We ignore here short-living particles.

It has now been discovered that electrons whiz around the atomic nucleus at a tremendous rate so that they do not collapse into the nucleus, which is almost at rest from the view of electron. The removal velocity for hydrogen is 1026 km/s. You can describe that with a high speed rotating fan. If you have a picture of it, the rotor blades will appear blurred, but the rotor axis is clearly visible. So it seems reasonable to deal with the around buzzing electrons with blur and quantum effect, but not in the description of the core structures. It therefore seems sensible to describe the many electrons rushing around the atomic nucleus with the means of quantum mechanics in order to gain at least an overview of the conditions in the electron shell. But exactly this quantum mechanical model was also used for the description of the atomic nucleus. Quantum mechanical statements are probable statements and no causal conclusions can be drawn from them, as we have stated above. And exactly this procedure is practiced in the standard model of cosmology. Since this model serves no scientific purpose, one can go beyond this logical barrier. The standard model of particle physics has the task to explain the ominous big bang, the creation, nothing else, not to find the cause of the radioactive decay of the atoms, which would be a worthwhile physical aim. Instead, it should explain how matter arises out of nothing. We should now take a closer look.

5.1 Criticism of the Standard Model of Particle Physics

Can particle physics be true beyond the limits of perception? The question came to me when I read the book by Alexander Unzicker *The Higgs Fake* [5.06]. It addresses the somewhat suspicious discovery of the so-called 'God Particle' on the world's largest ever-built electric machine, the Large Hadron Collider, where they seek to explain how particles come to their mass. Here, obviously, there has been a change of meaning in the concept of mass, for in atomic physics we understand mass as the number of particles, as we can see from the unit *u*. Alexander Unzicker intuitively and emotionally rejects the results of particle physics and the reasoning seemed to me to have little foundation. The book remains too attached to the phenomenons, without the fundamental problem of clearly working out the necessary demarcation of physics from mathematics as a spiritual science and to address the shift of conceptual content within physics. However, physics is still an empirical science, even if the theorists would like to change that. As empirical science, it is bound to the current detection limits of the instruments. Jörn Bleck Neuhaus [5.07], a representative of particle physics, attributes that particle physics poses for a challenge to the intellect, because it has noticeably blown up the boundaries of classical physics. But what are these boundaries? The boundaries are where we used mathematical imagination instead of the experimental findings. This, Unzicker criticizes rightly in my opinion. The crossing of this boundaries is, in the philosophical sense, metaphysics. However I dare to doubt, whether the claim he raises in his book, to address young people, is fulfilled. They would have to have a very good background in physics, philosophy and logic in order to form a judgment independent of the general doctrine. So the reader wonders: How can highly qualified experts be so wrong in their statements, however if they serve no discernible political goal? After all, you can not put them all under general suspicion of the imposture. On the other hand, the high degree of specialization of today's disciplines entails the risk of errors and misinterpretations, especially since the object of research eludes completely even sensual observation.

Here I want to pursue the question: Which results does particle physics claim for itself and how do these stand to the research logic and which claim to truth can their statements derive from it? For this the book *Elementare Teilchen – Von den Atomen über das Standardmodell bis zum Higgs-Boson* by Jörn Bleck Neuhaus [5.07] seems to me to be the best one. Here a representative of particle physics himself comes to word who really tries understandably to explain the results of his Discipline to the reader. For questions of research logic, please refer to section 2.3. Deductive and Inductive Logic in Research

5.2 Particle Physics and Cognitive Logic

Is it possible to apply the principle of falsification in particle physics, even though it is impossible for the outsider to understand the experiments and also to specify demarcation criteria? Both relativity theory and quantum theory emerged in the period when the Viennese circle of positivism had an influence on science. One of these representatives was Ernst Mach, whom Albert Einstein met at Prague University, where he had held a chair for three semesters, but his ambitions with regard to teaching were kept within very narrow limits. Einstein later claimed that Mach's mechanics had deeply impressed him and that Mach's example of critical thinking was the key to the theory of relativity. That is rather to be doubted, since his favorite philosopher Arthur Schopenhauer already anticipated the principle of relativity in 1818 in his book *The World as Will and Imagination* [5.08], Einstein published the special theory of relativity in 1905 and it was only in 1911 that he met Mach in Prague. This alludes to Mach's principle. The latter says that there is no absolute space as Newton understood. Movements of a body can only be recognized in relation to movements of other bodies. That says nothing other than that each observer brings his own frame of reference and the space is a mathematical concept in the mind of the observer. But this principle does not recognize the importance of a force field as part of a mass on the movement of the other body, which the observer can perceive as an acceleration force. The flea sees that the tail wags the dog if he does not feel the forces pulling on him. Observation is understood here as visual observation. That's why Einstein spoke of inertial systems. Only the inertia is connected with the mass particles and not with the space, a lot of independent characteristics like length, width and height. There is no physical space, but a particle-filled volume that is related to the concept of space, otherwise volumes could not be measured.

At that time science was philosophically very strongly influenced by positivism. Positivism stands for a position according to which the natural sciences should limit themselves to researching the observable facts ("positives"). Hume is regarded as the founder. The name *Positivism* itself goes back to Comte, who formulated the Three-Stages Law of Scientific Development: [5.09]

1. Theological Level: Real appearances are traced back to the working of one God or several Gods
2. Metaphysical level: comprehension of the general nature of things.
3. Positive level (highest): description of observed facts.

Thus positivism for cosmology indirectly demands a geocentric view of the world and goes back to positions before Galileo, but our actual view is heliocentric. Although the observer is included in the experiment, the imaging laws and other characteristics of the observer are neglected in the description of facts. On the contrary, both relativity theory and quantum theory take refuge in metaphysical regions beyond observable facts. While the theory of relativity considers observer and object as equivalent, without considering the energy balance in their movement, quantum theory does not differentiate between individual events and a multitude of events, which manifests itself in both the contradiction of wave-particle dualism and the linking of statistical statements and causal conclusions. This is a logical sin, which is often found in everyday life too. In order to avoid this dilemma, the particle physicists tried to describe all phenomena in a particle image, since they have recognized that they can not describe individual particles in a wave picture. We would not get stable particles from wave superpositions. However, everything discovered by high-energy particle physics is unstable.

The difficulty with particle physics lies in finding one or more suitable logical demarcation criteria, which naturally have to be different from the theory of relativity, and it requires a set of sentences to be investigated that particle physics claims for itself. From this, the inductively produced sentences must be filtered out and these must then be brought into conflict with other already recognized sentences. Particle physics is governed by the idea of symmetry, which is expressed in some of their sentences. Jörn Bleck-Neuhaus summarizes the current state of knowledge of elementary particle physics in 12 sentences:

1. *An elementary particle shows neither a finite spatial extent nor an internal structure.*
2. *There are only a few basic types of elementary particles. These are 2 varieties of fermions and 3 varieties of bosons.*
3. *Elementary particles can have angular momentum without spinning and being magnetic without a current flowing.*
4. *Elementary particles can be created and destroyed.*
5. *Every particle has an antiparticle.*
6. *Elementary particles of the same variety are indistinguishable.*
7. *The elementary act of electromagnetic interaction is the emission or absorption of a photon. The electrostatic potential also arises from this.*
8. *Elementary particles also produce measurable effects from 'unphysical' states in which they are unobservable (virtual states).*
9. *Each of the four basic forces of nature comes about through the exchange of elementary particles in virtual states.*
10. *There is an exact pictorial language for the interaction processes.*
11. *The four conservation laws of classical physics apply, but the mirror symmetries of classical physics are broken.*
12. *The particles can carry other types of charge, which can be partly converted into each other. This makes it unclear how many types of particles need to be counted as different.*

We now want to examine these sentences in order to find demarcation criteria. However, before we can deal with the individual sentences, some general remarks must be made. The first thing to do is to look at the initial situation and then discuss the experimental situation of particle physics.

5.2. 1 The Initial Situation and the Development of Quantum Mechanics

The interpretation of the optical spectral lines was not possible with classical physics, as it was around 1900. One could explain neither photo effect nor thermal radiation. Since the introduction of quantum mechanics was a huge development impulse. However, it was bound to the observations by means of a spectrograph. A spectrograph can not observe particles. So you have to imagine the path a light pulse travels through an optical medium as a wave. The idea of quantum mechanics is now the following:

Imagine, we have to cross a square, where many people seemingly aimlessly running back and forth. In order not to collide with the people, we would be forced to constantly dodge them right or left. We would not cover a straight path, but describe a more or less wavy curve with a random amplitude. The same thing would happen if we wanted to cross the square perpendicular to our first path. We would not even have to cross the square ourselves, a message that would be passed from mouth to mouth would take such a wave-like path.

You remember having read something like this before? This idea is important. Now we can transform this idea to the level of the atoms and call these directions, probability amplitude. The square of the absolute value of the function is called probability density. However, the inventors of quantum mechanics now explain that the movement of a free particle would accomplish this wave-like motion. In truth, however, the particle is not free at all, but moves in that way because there are enough other particles them have to dodge. We can therefore assume that a particle is, with a certain probability, at the intersection of two mutually perpendicular wave functions on its way. But that says nothing at all about the individual particle itself. In addition we can assume that the probability amplitude depends on which energy the place is crossed over. If somebody crosses the square with a motorcycle, all will make room as soon as they recognize the peril, if they do not want to run the risk of getting hurt. In the field of atoms, the injury means ionization. It does not seem to make sense to rely on quantum mechanics in high-energy particle physics. Here the concept of quantum mechanics is perverted. Obviously, energy contains a demarcation criterion for quantum mechanics. While quantum mechanics describes the probable behavior of particles, high-energy physics seeks exceptions in behavior, believing that every exception is a new particle, and subsequently making it the rule.

In classical physics, the particle has concrete coordinates, a direction, a mass and a velocity, and thus also a stability. Quantum mechanics says about a particle only that at the intersection of two paths there is a probability that there is a particle to be found there and if you integrate over the paths over an infinite range of values, then you get the certainty that at the intersection is a particle, but where the intersection is, the definition says nothing about it. You can not be sure because this integration is practically impossible. At the level of quantum mechanics, the term 'particle' is therefore completely indefinite, which is why they mistakenly speak of the wave character of a particle. You can not represent a particle through a wave packet because it would not be stable. This has nothing to do with the individual fate of the particle itself, but with the nature of the description. It is like photographing the particle with a too long exposure time. The picture is then correspondingly out of focus. Now every dynamic variable from classical mechanics in quantum mechanics is assigned a particular linear mathematical operator that acts on the wave function, assuming that the same identity relationships exist between the linear operators as between the sizes in classical mechanics. For example, the impulse is represented by the operator $-i\hbar$, which acts on the wave function... It is an imaginary operator in one spatial direction. Okay, that is still to understand mathematically. But what does that mean physically? Do we expose a particle existing in a spatial element probably to an effect? Do we hit the particle at all? We strike the volume element rather blindly and therefore we can not make any statement for a single particle at all. At most if we have done this many times, there is a possibility that we have met some particles. How high the hit rate is, remains unknown.

A model is used that gives good results for general statements, where many particles are involved in transmitting the effect *h* of an impulse, but it fails completely where statements are needed for an individual particle. **This contradiction is cultivated in quantum mechanics by speaking of the wave-particle dualism.** In addition, quantum mechanics can only be used for low-energy particles, i.e. if the energy of the considered particle is comparable to the energy of the neighboring particles. That changes total as we look at high-energy particles. For then the particle will fly straight through and the neighboring particles will suffer injury (ionization) if they do not dodge in time. We then obtain the corresponding cloud chamber images.

5.2.2 The Experiments of Particle Physics

The high-energy experiments of particle physics at the **L**arge **H**adron **C**ollider (LHC) are so complex today that they are incomprehensible by independent sources. The LHC is a gigantic circular particle accelerator with 27 kilometers of circumference, which is located in about 100 meters depth in the border area of Switzerland and France near Geneva. Around 9300 electromagnets generate a magnetic field with a flux density of 8.33 Tesla. (For comparison: The flux density of the magnetic field of the Earth is about 40 micro Tesla) The energy requirement of the LHC is around 120 megawatts - comparable to the electricity demand of the city of Geneva. In the magnetic field, an ion current and a proton current are accelerated to a velocity near light speed in opposite directions in order then to collide in a detector chamber. Astrophysicists stubbornly deny the existence of electric currents in space, but they have no qualms about using electrical energy to prove their electricity-free theory. It all looks like the miracle of an Immaculate Conception. Another curiosity is the fact that they believe that they can use crash tests to clarify the structure of matter. Transferring this idea to a technical level would mean that you could run crash tests with cars of the competition to explore the blueprint of these vehicles ...

Allegedly, a new particle is said to have been discovered during the experiments at the LHC. In July 2012, the physicists at CERN announced that they had found very clear evidence of the Higgs boson. The Higgs boson belongs to a theory proposed in the 1960s, according to which all elementary particles, except the Higgs boson itself, receive their mass only through the interaction with the ubiquitous Higgs field. The long-sought particle, which should decay into other particles after 10^{-20} seconds, would be 133 times as heavy as a proton at rest. Don't we already know from mass spectroscopy that a field belongs to every mass? What a particle should tell us, that actually doesn't exist, that produces mass from a mass-less field, that can not exist at all and that then transfers this mass to all other mass-less particles?

As late as October of the same year, the Nobel Prize Committee awarded the prize to the two researchers François Englert and Peter Higgs, stating

> *»...for the theoretical discovery of a mechanism that helps to understand the origin of the mass of subatomic particles, and most recently the discovery of the predicted fundamental particle in the Atlas and CMS experiments at CERN's Large Hadron Collider.«*

Note the phrase 'theoretical discovery'! (See section 2.2.2 Mathematics would not be a Concept, but Objective Reality) We could also say: "the discovery of alternative facts". What was confirmed? A very rare random phenomenon in the detector has been associated with an idea, that is very doubtful, and that nobody outside the research team of the LHC, can ever verify. We can not shake off the impression, that with the Nobel Prize the nonsensical economic expenses of this experiment, had to be justified. If they associate this Higgs particle with the overall existing mass, shouldn't there be expected, mass events of this type finding must be the most of at least, shouldn't it? I call this type of science the justification of a belief or a justification-science.

5.2.3. The Discussion of Sentences about Particles

Now we want to comment on the statements made by Bleck-Neuhaus in detail, and find demarcation criteria that divorce physics from metaphysics. As emphasized by Bleck-Neuhaus, theorems of particle physics are based on the contradictory duality principle of wave and particle.

1. *An elementary particle does not reveal either a finite spatial extent or an internal structure.*

In common usage, a particle is a spatially delimited piece of matter. A particle therefore has a clearly recognizable boundary as well as a mass, a charge and a force field. In daily life, the mass is still referred to a reference mass, the original kilogram. At atomic scale, the reference mass is the twelfth part of the carbon isotope ^{12}C. Everything we know about the mass is supplied to us by the mass spectrometer and it is electrically operated. **We obtain the mass as the ratio, of the product of charge multiplied with magnetic field strength divided the ion cyclotron frequency.** This particle definition has been very successful. She brought the explanation for the entire spectrum of elements with their isotopes by means of the particles electron, proton and neutron. However, the radioactivity of the atoms showed, that at least the neutron is not elemental because it turns into a hydrogen atom outside the nucleus with a half-life of 12 minutes. Even within the atomic nucleus, a certain ratio of protons to neutrons seems to be necessary for an atom to be stable. But maybe the neutron does not exist in the stable atomic nucleus at all. In the case of a hypothetical neutron surplus, the atom is radioactive. It converts neutrons into protons by electron emission until a new equilibrium state is reached. In the case of a hypothetical neutron deficiency, electrons are trapped by the nucleus until the equilibrium state is reached. However, looking at the half-lives between the individual nuclear transformations, we find half value times between seconds and centuries, without any tendency to be derived. We have no explanation for that. If we understood the inside of the atom, we could undoubtedly answer that question.

If you ask about the boundaries of an elementary particle, the answer is not very reliable, because the measurement methods are different and if you ask for its mass, you should not accelerate the particle. What do they actually want to measure in the particle accelerator? The electron radius was determined by the cross-section of about 3 fm (1 fm = 10^{-15} m), which agrees well with the classically calculated electron radius of 2.8 fm. Substituting the electron mass with the proton mass in the formula for the classical electron radius, you obtain for the proton radius by 3 orders of magnitude smaller value then measured. However, the proton radius is only about three times smaller than the electron radius, namely 0.8 fm. The neutron radius should be about 1.1 fm. [5.10] An older source states 5.7 fm [5.11]. WIKIPEDIA calls 0.85 fm. The measurements are based on scattering experiments. On average, you obtain about the electron radius with a fairly high standard deviation. The more energetic the scattered particles are, the smaller the target appears, which is due to the effect of the force field. Even Descartes did not regard a particle in its vortex theory as independent of its force field, which can always expand in dependence of its neighbors. For a long time, particle physicists believed that fragments of the atomic nucleus with a short lifetime were independent particles. In the meantime, however, the particle zoo has been restricted to a few basic types.

Demarcation criterion: When we speak of a particle, we must be able to distinguish it from its environment. I consciously avoid the term mass. Even the term 'form' is not justified here, since form can only be grasped by the sense of sight. Thus it also has charge and a surrounding force field, that prevents from entering an other particle. But the assumption that an empty space exists between the particles obviously can not be maintained. Even if we can not see a force field, forces can be measured, which is why they are material. In other words, particles fill up the available volume. It is the question of external pressure how big the particle appears. We can agree with the above sentence.

2. *There are only a few basic types of elementary particles. These are 2 varieties of fermions and 3 varieties of bosons.*

Particle physicists classify elementary particles by their spin. While fermions are assigned the spin ½ as the basic building blocks of matter, to which the electron and the proton belong,. The bosons to which is counted the photon bear the forces. They are assigned Spin 1. The fermions are divided into quarks and leptons. These quarks are theoretical constructs that are supposed to exist insides neutrons and protons. It is strange that only the leptons of the fermions are observable, for which 6 species are counted, including the three neutrino species whose existence is very doubtful. There remain only three observable energy states of the electron e, μ and τ, of which only e is stable. μ has a lifetime of microseconds and τ has a lifetime of less than a picoseconds.

Further, it is also strange to call the bosons particles, since they are quanta of action of forces that can only be transmitted in a force field. These forces occur only at the phase transition on a less penetrable phase, without particles would accumulate there. Even though they are quantized, effects do not leave mass accumulation behind. The bosons are assigned the spin 1.

What is the spin?

> The whole particle physics builds on the spin concept. But the spin has caused at least as much confusion as the wave-particle duality. To understand the spin concept, we have to look where it came from. The magnetic field shows that the spectral lines have a fine structure. Each line splits into three and more lines under the effect of the magnetic field. So there must be a feature on the electron that only becomes visible under the influence of the field. In 1921 Stern and Gerlach sent silver atoms through an inhomogeneous magnetic field [5.12]. Because electrons themselves have a magnetic field, they experience a force in the inhomogeneous magnetic field and are deflected. Classically, it is now expected that the axes of the small magnets can show in any spatial directions and the atoms fill the entire area when hitting the screen. In reality, however, only two bands were observed, as if there were only two possible adjustments. Because silver atoms have only one single valence electron in the s-sub-shell, and therefore have no orbital angular momentum, and thus no magnetic moment caused thereby too. The only possibility is that the electrons themselves have an "intrinsic angular momentum" and an associated magnetic

moment.
Another magneto-mechanical effect is the Einstein-De-Haas effect [5.13]. It shows that the electrons in a material sample align themselves to the field of a current-carrying inductor. Consequently, they must act like rotatable elementary magnets. From this effect results the mechanical moment of the electron. Consequently, they must act like rotatable elementary magnets. From this effect results the mechanical moment of the electron. The magnetic moment, once derived from Bohr, and in the second case as a mechanical moment, provides the remarkable difference of exactly ½. Since Schpolski noted that the torsion of the sample in the magnetic field in the Einstein-De-Haas effect was half a turn and the derivative of the Bohr magneton was calculated with one complete turn, there is the suspicion that in comparison of theoretical derivation of Bohr's magneton and Einstein-De-Haas this difference was not considered. Because in [5.13] only a ten percent deviation from the predicted value was given. Instead, this factor got the name spin, although obviously it is just a misunderstanding, and has nothing to do with rotation, but follows from the unequal calculation bases for the moments. Then the spin was introduced as a hypothesis to bridge the discrepancy between the Bohr magneton and the magnetic moment from the Einstein-De-Haas effect. Schpolski writes in his atomic physics Bd.II §197: *»The idea of the electron spin and the associated peculiarities were introduced as a hypothesis, but it became clear that the existence of the spin and all its properties from the Dirac equation of quantum mechanics, the demands of the theory of relativity does justice.«*
This is undoubtedly a mistake, since the spin of ½, was entered as a prerequisite in the Dirac equations. These are four coupled differential equations for each coordinate in space-time. From a model you can not get more knowledge, than you put into it. Consequently, the spin is a factor that establishes the relationship between the Bohr magneton and the mechanical angular momentum for half a turn. Instead of clarifying the question of why this discrepancy between the two derivatives of the magnetic moment came about, they simply put forward a theory, that provided more confusion than clarification. Thus, the spin should have all the properties of a classical mechanical angular momentum except that it is caused by the rotational motion of a mass, and in quantum mechanical formalism it is a simple number, although physically a vector would be expected according to its designation. As long as electrons are considered as small magnetic dipoles, this is incomprehensible. However, this changes if one assumes that electrons do not have a dipole field, but a vortex field, as we have tacitly assumed in section 4.4.2 in the transition from a conductor to a single electron. Then we would not need to distinguish between magnetic and mechanical moment any more. Evidence of the vortex field of the electron was recently obtained by J. de Climont [5.14]. What they already properly not understood at the electron, they transferred it to the proton, whose magnetic moment is about 660 times smaller.

To the proton is attributed the spin ½ since 1928, because an anomaly in the specific heat of hydrogen gas could not be explained otherwise [5.15]. The proton spin is said to have been detected first in 2011 on a single proton. The original paper is not freely available.

Unlike the half-integer spin of the leptons, the integer spin of the photon (quantum of light) is said to arise from the long known existence of electromagnetic waves with circular polarization. In fact, all that means is that the magnetic moment coincides with Bohr's magneton, which raises the question of how to determine the magnetic moment of a photon. According to Sommerfeld [5.16], circularly polarized light can be produced by twice total reflection on glass prisms. Normally, light is not circularly polarized. The Maxwell equations also do not form a circular polarization. The operator curl has a different function there, it is a circulating integral. However, the rotation of a wave is something completely different from the rotation of a particle, because a wave always consists of many coupled particles carrying an effect. Opposite to claiming that direct experimental evidence of photon spin in 1936 was based on the rotational motion of a macroscopic object after interacting with photons, it can be argued that a mill wheel also rotates without the water flowing underneath forming a vortex. There are a lot of technical applications where a straightforward motion is translated into a rotary motion.

Demarcation criterion: Bosons are not particles, they are action pulses, which explains their extremely short lifetimes. Spin quantum numbers are not physically explainable. They are probably due to a misunderstanding of the basis of calculation (full revolution, half revolution). This would mean that the particle system built up on the spin would be void.

3. *Elementary particles can have an angular momentum without turning and being magnetic without a current flowing.*

At the beginning we discussed demarcation criteria for the distinction between physics and metaphysics. Here it is actually claimed that elementary particles have an impulse without movement. The momentum is the product of mass and velocity. The angular momentum is defined as the cross product of radius and attacking force. The force is defined as the product of mass and acceleration. An impulse can only come from a mass. The effect of the pulse itself, however, has no mass. It is the effect of a moving mass on a resting mass. Anyone would tell so who thumped to his finger with a hammer. The effect is visible without the finger enlarging about the mass of the hammer, even though the swelling may then be considerable. The hammer stays the same when you make a second hit without doubling themselves. Photons are like hammer blows, they are not particles, but they produce effects at phase boundaries. Consequently, the use of an activity quantum to characterize it is justified. Photons, like all bosons, are like hammer blows, but calling them particles is a foolish act[5]) and why they should have a spin is incomprehensible. The action quantum (or Planck constant) is defined as the product of energy and time and not as an angular momentum.

Measurement units of the action quantum: erg s → cm² g/s

There is one action quantum per revolution. It is denoted by $\hbar = h/2\pi$. When describing an orbital angular momentum, it has a spin axis and an amount. This means that amount and direction are variable. Only in a field it comes to alignment. The amount is h and for the direction you have the axis through the full circle 2π, the projection on the field direction results according to the quantum mechanics only 7 values.

Mathematicians can not get used to the fact that physical quantities always have a unit of measure. The spin is specified exactly in $n \cdot \hbar = n \cdot h/$

5 It is said about the citizens of Schilda that they forgot the windows when building their town hall. Sitting in the dark at their council meetings, a clever shield citizen came up with the idea of bringing the photons together on the light-flooded street and carrying them in sacks to the town hall. If I recall correctly, the idea came from a certain Albert. Citizens of Schilda is the German equivalent to the Wise Men of Gotham

2π per radian, that is to say in relation to the full circle. The unit is called the effect per revolution. If you give the spin to ½ $\hbar$, that means half the effect per revolution. But in which direction? In addition, h was postulated as the smallest quantum of effect that we could not divide anymore. Stephen Hawking [5.17] used a curious explanation for the spin in his book *A Brief History of Time* using an arrow analogy to illustrate the spin:

> *»A particle with the spin 0 is a point: it looks the same from all directions. A particle with spin 1 is like an arrow: it looks different from different directions. Only with one complete revolution (360 degrees) does the particle look the same again. A particle with the spin 2 is like an arrow with a tip at each end. It looks the same after half a turn (180 degrees). Similarly, higher spin particles look the same again when you spin through smaller fractions of a full turn. In addition, there are particles that do not look the same again after one revolution: Two complete revolutions are required! The spin of such particles is given as ½.«*

Since there the logic is already interrupted again. To argue with the appearance of the spin of particles below the limit of detection and, moreover, in the above manner, is evidence of either imbecility or stultification. In any case, particles must have a mass to spin. If they have a mass, then they also have a charge and a magnetic moment and they have a thermal energy, which contradicts the above claim of their immobility.

Demarcation criterion: Light quanta do not have a double character. They are effects, not particles. Quantum mechanics is misinterpreted, it does not apply to individual particles. It can only make statistical statements. The Copenhagen interpretation of quantum mechanics, however, denies a statistical interpretation, which makes the matter even more problematic.

4. *Elementary particles can be created and destroyed.*

Mass and energy can not be created or destroyed in our experience. But they can seemingly be transformed into each other and elude our immediate observation. In fact, it sometimes looks as if individual particles disappear or re-emerge. However, since we can not observe these particles ourselves, but only their effect on the environment, we can only say that the effects at phase boundaries can be visible and otherwise remain invisible, which is also macroscopically plausible. The thumb surface is the phase boundary at which the effect of the hammer blow unfolds. Since particles can be either positively or negatively charged, the abandoned, originally neutral environment of the particle will always show the counteracted charge and appear as a positively or negatively charged hole. If a particle is torn out of its proper place, a 'hole' remains behind, giving the impression that a particle with the opposite charge can be produced. If it returns to such a hole, it gives the impression of being destroyed. Holes are basically unstable.

Demarcation criterion: If a particle were destroyed, its energy and mass would be destroyed as well. The energy equation would be overridden and the door opened to the belief in creation. This contradicts that in section 3.3. defined sentence of the conservation of matter.

5. *There are antiparticles to particles*

There are many things that can not really be explained, crossing the physical detection limit. For inexplicable appearances, empty terms such as black hole, dark matter or antimatter have been introduced. The term 'antiparticle' is derived from antimatter. Antimatter grew out of the belief in the mirrored symmetry of the world. Matter as the philosophical category of materialism is defined as everything that exists outside of our consciousness, hence our awareness should consist of antimatter, would be an obvious conclusion. So but Matter has no plurality, as do the concepts universe and infinity. Consequently, the antiparticle derived from antimatter has no demonstrable relevance, especially since philosophers could not even agree on a definition of the concept of matter among themselves. Immanuel Kant would call the term 'antiparticle' as an empty term. As happens with all empty concepts, contradictory meaning contents are interpreted into them.

Demarcation criterion: The symmetry principle is not a fundamental principle of nature, which is why thesis 5 is invalid. There is no antiphysics and no theory about anti-energy and anti-waves with anti-speed and anti-time. And there is no measure to distinguish particles from antiparticles. But there are effects in the micro-world, that we do currently understand nothing, that we provide with such joker terms.

6. *Elementary particles of the same variety are indistinguishable.*

If elementary particles of the same variety are indistinguishable, then they are not countable either. In other words, quantum mechanics has no way of describing a single particle. This statement collides with the atomic models, which of course has countable elementary particles. Even their residence probabilities are distinguishable according to orbital theory.

Demarcation criterion: Quantum mechanics is not applicable on single particles

7. *The elementary act of electromagnetic interaction is the emission or absorption of a photon. The electrostatic potential also arises.*

Photons are effects of the electromagnetic force field due to the movement of electrically charged particles, but themselves no particles, because they can not have a charge. If they have no charge, they can not have electrostatic potential. In this respect, the elementary act can not be the emission or absorption of a photon, but only the consequence of charge separation and recombination of charge carriers.

As photons (wave trains) can act on particle bonds by virtue of their resonances, photons are able to dissolve particles out of their structures, which is demonstrated particularly impressively in material evaporation by means of a laser. The thermal effect shows that even low-energy absorbed thermal radiation at phase transitions is able to generate and maintain charge separation. An electrostatic potential always requires a double layer of different phases that maintains charge separation. Batteries are not operated exclusively by light. Other forms of energy are also used, inclusive mechanical energy.

Demarcation criterion: Effects are no particles. The electrostatic potential is due to the opposite charges of protons and electrons. The electromagnetic interaction arises from the movement of a charge in the stationary force field.

8. *Elementary particles also produce measurable effects from "unphysical" states in which they are unobservable (virtual states).*

It is unclear what an unphysical state should be and what it has to do with physics. The term 'virtual' we know from the optics. It refers to the mirror image of an object that we believe to recognize behind the mirror by its reflection. Pulses produce effects at phase boundaries without being able to observe directly their transmission in a medium. Particle physics does not use a material force field to interpret its results but an abstract empty space. Therefore, it can not explain the transport of effects. However, since physics is an empirical science of measurement, it can not make reliable statements about unobservable states. Elementary particles are only indirectly observable, not directly measurable. We can only perceive them indirectly through their effects. As a result, misinterpretations are easily possible. Each gauge is orders of magnitude larger than the elementary particle to be measured. Consequently, the interference on the object to be measured is large, which affects the measurement error.

Demarcation criterion: This statement is completely unacceptable. It testifies to the lack of understanding of the elementary processes.

9. *Each of the four basic forces of nature comes about through the exchange of elementary particles in virtual states.*

If meant under unobservable states virtual conditions, then who should recognize that such states exist. Immanuel Kant described a term without intuition as empty. For 'virtual' you can therefore set 'unknown' or say that you do not know how the exchange should come about. The use of the concept of the god particle hits it, since the empty term 'God' in history has always been used for the inexplicable to fill it with beliefs.

The mainstream physics distinguishes two short-range nuclear forces acting only within the nucleus as well as the electric force and gravity outside the nucleus with infinite range. However, it has been shown that the electrical force can be combined with the nuclear forces in a common theory. Only gravity does not seem to fit in because it is not shieldable. However, the shielding is guaranteed by electrically conductive shield materials and is the work of free electrons. The shield is placed on Earth voltage. The shielding is the work of free electrons. Gravitation has nothing to do with free electrons. It seems illogical that we want to distinguish several types of forces, although there is only one unit of measure for forces. However, since force has a direction, it makes more sense to differentiate forces according to directions. The force is a vector, given by a direction and an amount with a unit of measure. Thereafter, you can not distinguish four basic forces, but each direction of force can be composed of three basic directions with their amounts.

Demarcation criterion: This statement is completely unacceptable. It testifies to the lack of understanding of the elementary processes.

10. *There is an exact pictorial language for the interaction processes.*

Although quantum mechanics, with its dualism of wave and particle, deliberately violates the laws of intuition and builds upon the insights discussed here, it breaks with the imagery of this principle by using the Feynmann graphs to illustrate believed physical processes. Only such processes were not really observed in nature .

Demarcation criterion: Due to the rejection of the previous statements, this statement is doubtful.

11. T*he four conservation laws of classical physics apply, but the mirror symmetry of classical physics is broken*.

The four conservation laws mean energy, mass, momentum and charge. What is meant is the symmetry of the Hamiltonian equations of classical mechanics. This sentence contradicts statement 4 about the disappearance and emergence of elementary particles.

The symmetry is already broken with the second law of thermodynamics and belongs to classical physics. It is also strange to want to derive these conservation laws from symmetries, because symmetry arises from the watching angle. and so the cause is exchanged with its effect. By contrast, conservation laws are basic assumptions of physics based on experience. Another ancient assumption is that of the unity of opposites, which is expressed physically in the opposition of charges, which neutralize themselves.

Reflections in space and time are virtual. Virtual means here, we believe to see something that does in reality not exist. Real nature is self-similar, symmetries on surfaces are part of it. This means that similar structures can be repeated on different scales. For example, a negative charge at a thousand-eight hundred times mass finds its positive equivalent. This is not a charge symmetry, but the starting point for the diversity of nature surrounding us. The confusion between self-similarity and symmetry stems from the fact that theoretical physics at the beginning of twentieth century, had reckon so the observed picture for reality, but it could not deduce the reality from two different pictures of an object. It was usual to follow the subjective idealism and the idea of "The World as Will and Intuition". Arthur Schopenhauer's critique of Immanuel Kant's "Critique of Pure Reason" had a long-lasting effect.

The fact that you can watch movies backwards does not mean that time is symmetrical. Our time refers to the clocked flow of energy from the sun to the earth, even if you measure it in oscillations of a Cs line today. The energy flow has only one direction, following the potential difference for compensation and with increasing entropy. Who wants to reverse that? Even the often-used argument that causality ceases to apply in the microcosm is just an evasive defense. Those who claim this, must declare the border between the micro world and the real world.

Demarcation criterion: The statement contradicts the Big Bang and is also contradictory in itself because of the adherence to the symmetry.

12. The particles can carry other types of charge, which can be partly converted into each other. This makes it unclear how many types of particles need to be counted as different.

When new kinds of charges are invented in elementary particle physics, only our imagination limits to doing this. These statements prove that the transition to metaphysics is complete and any further commentary is superfluous.

Demarcation criterion: The statement must be rejected completely. There is only electrical charge separation as the cause of the forces observed.

5.3 Conclusion

From the engineer's point of view, only the first statement is a valid statement: *An elementary particle does not reveal either a finite spatial extent or an internal structure*. All other statements by Bleck-Neuhaus on particle physics, it must be noted, are based on the one hand on the idea of the duality of wave and particle, and on the other hand they are carred of the belief in the symmetry of the world beyond the limits of the measurable. A logical mistake is the distinction between magnetic moment and spin and the fact that the spin was the basis for the qualitative differentiation of the particles. This assigns a physical quality to a simple possibly even erroneous quantity. There also is a problem with conservation laws in particle physics.

Physically, particle physics offers nothing that would not have been known before. the use of hypothetical particles with auxiliary charges also fails to answer the question of why one type of atom is stable and other atoms are radioactive.

Both the idea of the duality of wave and particle as well as the belief in the symmetry of the world are not tenable on closer inspection and physics has to delineate against it. On the one hand, I can design a wave pattern of nature and on the other a particle picture, but I can not say that these pictures are the nature itself. It is completely misleading to use the images and to draw inductive conclusions for an unobservable existence. Who can really judge if something unobservable exists? One can assume at most one existence behind an observable phenomenon. But to make this assumption to the basis of further speculation, to objectify it, is a logical sin. That must inevitably lead to contradictions.

In the second case, symmetry, we need certain points of observation to detect symmetries. From a different perspective, however, no symmetries are observed. Symmetry is thus dependent on the view point with the exception of the spherical symmetry. This is where the philosophy of Arthur Schopenhauer, who declares the principle of relativity between observer and observed object to which both the Copenhagen school of Heisenberg [5.18] and Albert Einstein have sat up, because they did not question the imaging laws that affect the observed image. None of our senses is easier to fool than our sense of sight. The ancient Indians had their own goddess, called Maya, the power of deception.

The claim to conduct basic research proves to be only an excuse here, and experiments merely serve to pseudo scientifically justify the preconceived belief in the Big Bang.

5.4. The Droplet Model of the Atomic Nucleus

We have recognized that in the microcosm causality is not abolished, but that the quantum-mechanical method of view eliminates causality due to its statistical approach to data reduction and after the standard model of particle physics has proved incapable, also because of the misinterpreting of the difference of both calculation methods of the magnetic moment as spin. To give a physical answer to why the atomic nucleus is stable in certain isotopes and unstable in others, and what the nuclear forces really are, we must return to older concepts. How should one imagine the inside of an atomic nucleus? If you look around in the literature [5.19], you find very different core models. The well-known models of the atomic nucleus show two opposite, strongly simplifying starting points:

- *Models of strong correlation:* The atomic nucleus is understood as a cluster of closely paired nucleons (eg, droplet model, alpha particle model by Wefelmeier, proposed in 1937, based on the assumption that nuclei are composed of a large number of alpha particles).
- *Models of independent particles:* The nucleons move relatively freely in the nucleus (Fermigas model, optical model, shell model, potential wall model).

The 'Droplet model' comes very close to the observed phenomena, which is why we also speak of a correlation model. It describes the atomic nucleus as a globular droplet of an electrically charged liquid. The basic idea was developed by George Gamow in 1935. It became known as the **Alpher-Bethe-Gamow theory** ("αβγ theory") and was the first theory of element formation in the early Universe when 1948, after the war, it was published titled as *The Origin of Chemical Elements* [5.20]. This work eventually stimulated Lemaître's big bang hypothesis [5.21]. Lemaître believed, that the world had emerged from a huge atomic nucleus that suddenly burst. This idea later transformed into an 'explosion from nowhere' that is supposed to represent the act of creation. Before 1936, however, Niels Bohr had already developed a droplet model (compound nuclear reaction as a possible mechanism of nuclear fusion reactions [5.22]). Lise Meitner and Otto Frisch used a droplet model in 1939 for the first explanation of nuclear fission and the released nuclear energy [5.23].

The spectral classes of the stars of our galaxy clearly show the origin of the chemical elements from the star-fire, as Gamow and Bohr foresaw. In addition, we have learned from the spectra of galaxies that hydrogen is by far the most abundant element in the intergalactic space, and that galaxies can be classified by their hydrogen content. I did not find helium in galaxy spectra, but it is in Star atmospheres of spectral classes O and B available [5.24]. Whether the merger of atomic nuclei in the stars will proceed as one currently imagines, I dare to doubt. But to this topic later.

5.4.1 The New Electromagnetic Droplet Model

The work of Carl Johnson where he concludes: »*The analysis of the exact NIST data seems to indicate that there is NO energy in the atomic nucleus, to explain the existence of any π-mesons, or the necessary binding energy of any neutrons, or any ultra-strong nuclear power or any neutrinos.*« [5.25], provided an opportunity to think afresh about a core model, If discrete neutrons actually existed in all nuclei, it seems logical, that at least some nuclei would naturally decay with releasing of one or more neutrons. That was not observed. Only by an external disturbance could be observed the release of neutrons, for example on beryllium or by nuclear fission on heavy atomic nuclei. Johnson argues:

> »*We can see the natural decay of tritium (hydrogen-3) with a half-life of*

12.33 years in helium-3 and a deconvolutional electron (β-particle, which is then recaptured as a circulating electron). This situation is clearly one in which exactly the same amount and number of objects are involved, three protons and three electrons, but some of the protons and electrons are (allegedly) bound together as neutrons in the nucleus. The laws of conservation of mass and energy certainly apply, so that a strict energy balance for this decay must show exactly the same total energy/mass before and after decay. In the initial situation, there should be a neutron that no longer exists in the final situation. The difference in the total energy (mass) contained in these nuclei of the two nuclei should therefore be the 0.78235 MeV of the binding energy of the one neutron within the tritium nucleus, which is no longer a neutron. However, using the accepted atomic mass NIST data, the difference in atomic masses is only 0.0000199578 atomic mass units (3.0160492779 - 3.0160293201) or 0.0185906 MeV. So this is the total amount of energy available to be released on decay while conserving energy. Since it has been shown experimentally that the escaping electron carries away 0.0185906 MeV kinetic energy, no energy is produced which suggests an initial neutron-bond energy of 0.78235 MeV. The energy responsible for this decay is particularly simple and particularly clear in confirming that the disappearance of mass according to the NIST numbers is essentially exactly explained by the kinetic energy of the escaping electron. NO possible neutron binding energy could have existed! « [5.30]

According to Johnson, the error analysis of this experimental data shows that for the escaping neutrino, less than 1 electron volt of energy would remain to escape. The classic solar model needs the neutrinos to explain the heat flow from the interior of the sun, which would be difficult to achieve with the amount of energy given above. On the other hand, it was already known in 1960 that in the solar corona temperatures of several million degrees prevail [5.26], but on the sun's surface less than 6000 degrees, which is not to explain the classic solar model. In addition, the many spectral lines of higher-order elements can not be explained.

Now let's design a new atomic model. Let's start by asking: which particles are stable? As we have already described, of all the particles with sufficient stability that have been described in the standard model, only the electron and the proton remain. Using the radii of electron and proton described earlier, we obtain the volume for an electron as a sphere 33 times larger than the proton. Whether the electron and the proton are really spherical, we can not know. The form is meaningless. The mass density of the proton is 2400 times higher than that of the electron. So the idea that electron and proton coexist in the nucleus is a little outlandish. Nevertheless in textbooks, in most cases the electron is shown to be small compared to the proton. The proton floats more in the negatively charged electron fluid because of the fact that the inner of the electron is field-free like we learned in 4.4.2. On the other hand, according to Maxwell's equations, we can assume that the electron is a doughnut-shaped vortex.

A neutron existing outside the atomic nucleus would then be a proton, which would be embedded in an electron. This neutron decays into a proton and an electron with a half-life of 12 minutes. The physicists' doctrine claims that this decay would create a neutrino, which would have to absorb the electron's recoil because of its energy distribution. E. W. Schpolski explains in his *Atomic Physics Bd.II* that the experiments and the calculations showed that a neutrino ionized no more than one pair of electrons per 500 km of airway [5.29]. But this means that the evidence is not experimentally possible, because no one can guarantee that on the long way not any other radioactive decay can happen. According to Johnson, the energy that passes to a neutrino would be so small that it would not even be enough for one ion pair. For nitrogen, the ionization energy is 14 eV. Since Johnson's data is accurate to a tenth of an eV, the existence of neutrinos in the atomic nucleus is impossible. The claimed resting energy is less than 2.2 eV according to WIKIPEDIA. That's not enough to ionize in a detector ones of the atoms.

Nevertheless, the Laboratorio Nazionali del Gran Sasso claims to have detected about 10 neutrinos per day at a depth of 1400 meters in Abruzzo. The neutrinos would have been detected by neutrino electron scattering in a 300 ton unsegmented liquid scintillator. Since May 2007, the scintillator BOREXINO has collected data. The aim of this detector is to detect direct solar neutrinos from the trapping of electrons in the radioactive isotope ^{7}Be with a half-life of 53 days and an ionization potential of 9.3 eV (!). Anyone who has ever registered nuclear decay with a scintillator can not imagine how to find something that should be a neutrino, because these amplified signals would have to be superimposed, and only the amplification effect provides enough reason for misinterpretations of the background. But the cost of such an experiment can not be justified if it turns out to be negative.

The real problem Pauli had was the quantum mechanical spin control. For Pauli, the spin was a quantum number that had to satisfy mathematical formalism. As we have already discussed, spin is only a factor between two different calculation bases for the magnetic moment. The torque of a body is a vector that lies in the axis of rotation of the body and gives two different orientations due to two different directions of rotation of the axis according to the right hand rule. The magnetic moment is also a vector. These are two forces that either add or subtract. After all, only one resultant force can be measured. The fact that there is such an orientation in electrons was first discovered by Stern and Gerlach [5.27] on silver atoms in 1922, and 5 years later the discovery of a spin was also reported on the hydrogen atom [5.29]. The electron spin was observed by the splitting of an electron beam in the magnetic field. The spin, however, is obviously a phantom created by a misunderstanding behind which the magnetic moment of the elementary particle hides, as discussed above.

The search for an electric dipole moment of the electron has so far been without positive findings [5.33][5.34]. The fact that there are electrical elementary charges, but not the magnetic counterpart, magnetic monopolies, has repeatedly led to attempts to remedy this asymmetry. So they supposed an electric dipole. In 2016, J. de Climont demonstrated in an experiment that electrons have no dipole field but a vortex field [5.35].

Since both after the energy balance of C. Johnson [5.30] and with the calculations of Schpolski, the neutrino predicted by Wolfgang Pauli in 1930 turned out to be a phantom, it is not to be expected in the atomic nucleus. The nucleus of the neutron is then nothing other than the proton, that has a mass comparable to a car of 1,860 kg, which emits 1 kg of combustion gas from the exhaust and is enveloped by that cloud. So what should take the recoil energy of the electron cloud that leaves the nucleus, if not the remaining proton? For example, when the proton is in the tritium nucleus, this recoil energy is sufficient to invade the remaining electron while the original host electron leaves the tritium nucleus. In other words, the proton leaves the one nuclear electron due to unknown external influences with a half-life of 12 years. When an electron leaves the nucleus, obviously only one nuclear electron with three protons remains, which is unusual and must now be discussed because it seems to be a special case. Normally, the number of protons that hosts a nuclear electron is no greater than 2, as we shall see. The magnetic moment is perpendicular to the vortex. For example, two magnetic vortices can have parallel or anti-parallel moments. According to the latest measurements, the magnetic moment of the proton is extremely small compared to the neutron. With three protons there is the possibility that the moments form a cyclic triangle or that a moment is countercyclical.

The whole theoretical particle physics is mysterious, especially because it is so completely incompatible with classic electrodynamics. So far, it has not led to any useful findings. In the field of nuclear fusion, it has been standing still for almost seventy years, although it obviously is also electromagnetic processes, which are based on Maxwell's equations, which have been so well proven in our everyday lives in almost all industrial applications. Ultimately, our entire industry relies on applications of electromagnetism, except for the generation of energy, which has been based on the combustion of fossil fuels since the first technical revolution.

Since the Maxwell equations describe electromagnetism and we only have two stable particles left out of all high-energy physics, a massive magnetic core and a fluid electron loop, then the whole standard model of particle physics will collapse and we should finally overcome the self-imposed limits of our thinking. Looking at the equations symmetrically tells us that there is a current circuit and a magnetic vortex that are mutually dependent. Maxwell says nothing about the size of the vertebrae.

Can't we think of the atom as a Tesla transformer, which is characterized by the fact that it has one winding with only a few turns and another winding with a large number of turns? The atomic shell forms an external current loop. The current loop in the neutron is three orders of magnitude smaller. As a result, the electron makes three orders of magnitude more turns in the neutron than in the atomic shell. However, in order to get into the higher orbit, the electron must first absorb energy from the environment in order to overcome a potential wall and not release energy in the form of a neutrino, which is, however, in contradiction to the current statements of the theory. Our experiments are always designed in such a way that they should prove a theory, never refute a theory as Karl Popper demanded.

A neutron is not completely electrically shielded from the outside and because of its magnetic moment it should also have a magnetic field. The solution of a potential field provides the Bessel function. This fact is completely ignored in nuclear physics. The consequence is the shell structure of the Atom shell. According to the Bessel function, there are also preferred trajectories on which particles in the potential field can move force-free regarding the center, as is assumed without explanation in Bohr's atomic model. But there the Lorentz force acts perpendicular to the center.

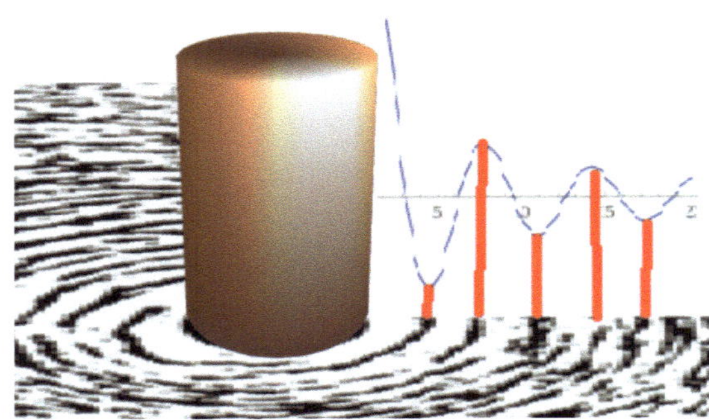

Figure 5.1: Magnetic potential field visualized by iron filings with Bessel function

Instead of assuming any quarks in the atomic nucleus, it is therefore obvious to assume Maxwell's equa-

tions not only for the radiation of electromagnetic waves but also for the structure of the atomic nucleus and to understand the smallest magnetic vortices as protons and the smallest electrical vortices as electrons.

This defines the role distribution in the model. The ferrite core symbolizes the proton and the wire winding the much lighter electron. Now we can build different models with ferrite core and wire loop. For the sake of simplicity, we will represent the proton as a red colored doughnut and the electron as a blue colored doughnut, and since the electron appears to be larger than the proton, we put a spherical shell around one, two or three protons.

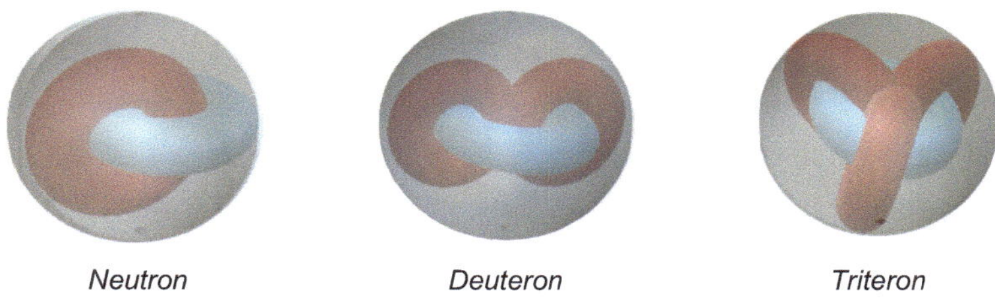

Neutron *Deuteron* *Triteron*

Figure 5.2: Core modules

We can build a neutron from a red doughnut and blue doughnut. However, we can also combine two or three red donuts with a blue doughnut and thus simulate a deuterium nucleus, hereinafter referred to as *Deuteron*, and a tritium nucleus. We simply call the latter *Triteron*.

We want to find out if an approach based on basic atomic structures to help us in the construction of atomic nuclei here. If, under rare circumstances, we naturally find unstable neutrons outside the atomic nucleus, why should they suddenly be stable in the nucleus? The standard answer is always, that's because of the nuclear forces.

But physically we can not differentiate between varieties of forces. Their range should be shorter than that of the electric forces, but where should these forces come from? Apart from the neutrons supposed to be found there, only protons are to be found in the nucleus. But they only have electromagnetic forces. The right answer could lie in the internal structure of the atomic nucleus, in a structure that is not visible but destroyed by high-energy bombardment. This idea you can found by Edwin Kaal [5.36] for example. Kaal makes responsibly electrostatic forces between proton and electron for the cohesion in the atomic nucleus and the nuclear field is supposed to influence the atomic shell.

If we look at the right core building block in Figure 5.2, we have three protons with one core electron. We cannot say whether it is a ^3H or a ^3He nucleus. The chemical properties are determined by the number of shell electrons. The magnetic moments of the three protons each form angles of 60 degrees with each other, canceling out the total momentum of the protons. As a total moment, only the moment of the enveloping core electron remains. Both ^2D and ^3He droplets can be considered as the magnetic building blocks for stable isotopes, as we will see below, while a neutron can occur only in radioactive isotopes. Deuterons can be used to build all atoms with an even number of masses. For an atomic nucleus with an odd mass number, we also need a triteron.

In the atomic nuclei, droplets of deuterons form into polyhedra as a result of the magnetic forces, or they are grouped around the droplet of a triteron, as the tables in section 5.4.2 shows, as Niels Bohr had already imagined nuclear fusion reactions as a compound nuclear reaction in 1936.[5.23]

In the following tables, the column Z contains the atomic number of the isotope and M the mass number of the nucleons, which is equal to the number of protons. To talk about the mass at the atomic level is a bit misleading, since the elementary particles are countable. Instead of the mass number one would have to speak more exactly of the mightiness. The term 'mightiness' comes from set theory and stands for the number of elements of a set. From the mass number, the idea of the mass of elementary particles is derived as a mystical property obtained by dividing energy by the square of the speed of light. We have already talked about the problem of the equivalence relationship of E and $m \cdot c^2$, which we know particles never reach. This property, which we identified as inertia in section 4.4.1 The Mass from a Macroscopic Point of View and recognized in 4.4.2 The Electrodynamic Mass of an Electron as being caused by the magnetic field, appears to us due to the motion of a charge as a magnetic field force.

The neutron number in the old model corresponds to the number of nuclear electrons that results from M-Z.

5.4.2 The Table of Construction of Selected Stable Isotopes

The basis of the droplet model is the nucleotides map of Hermann Ebert [5.38]. From this is derived the following table of stable isotopes. Three columns 'droplets' with stable isotopes of the respective element were shown, with column 'M' indicating the mass number of the isotope. Two more columns contain the number of deuterons and triterons per isotope. The column 'M / (M-Z)' is the ratio of positive nuclear charge to negative nuclear charge. For us to have a balanced charge ratio throughout the atom, this value must be about two. One nuclear electron and one shell electron each face a proton.

The lines of blue numbers in Table 5.1 each show only one stable isotope, while there are usually several stable isotopes of each element, especially as the number of protons grows. The significance of this we first see in the radioactive isotopes. It also shows that the ratio of positive nuclear charge to negative nuclear charge with increasing atomic weight is less than two, without the stability of the atom collapses immediately. However, if this value becomes greater than 2, it will cause immediate instability. This is not the case only in the case of a single Triteron droplet. For this reason you can assume that certain atomic nuclei inside themselves accept one Triteron core building block, around which the deuteron blocks group like magnets.

Stabile Isotope

Element	Z	M	Tri	Deu	M/(M-Z)	M	Tri	Deu	M/(M-Z)	M	Tri	Deu	M/(M-Z)
			Tröpfchen				Tröpfchen				Tröpfchen		
H	1	1				2		1	2				
He	2	3			3,00	4		2	2,00				
Li	3	6			2,00	7			1,75				
Be	4	9	1	3	1,80								
B	5	10		5	2,00	11	1	4	1,83				
C	6	12		6	2,00	13	1	5	1,86				
N	7	15	1	6	1,88	14		7	2,00				
O	8	16		8	2,00	17	1	7	1,89	18		9	1,80
F	9	19	1	8	1,90								
Ne	10	20		10	2,00	21	1	9	1,91	22		11	1,83
Na	11	23	1	10	1,92								
Mg	12	24		12	2,00	25	1	11	1,92	26		13	1,86
Al	13	27	1	12	1,93								
Si	14	28		14	2,00	29	1	13	1,93	30		15	1,88
P	15	31	1	14	1,94								
S	16	32		16	2,00	33	1	15	1,94	34		17	1,89
Cl	17	35	1	16	1,94	37	1	17	1,85				
Ar	18	36		18	2,00	38		19	1,90	40		20	1,82
K	19	39	1	18	1,95	41	1	19	1,86				
Ca	20	40		20	2,00	42		21	1,91	43	1	20	1,87
Sc	21	45	1	21	1,88								
Ti	22	46		23	1,92	47	1	22	1,88	48		24	1,85
V	23	51	1	24	1,82								
Cr	24	50		25	1,92	52		26	1,86	53	1	25	1,83
Mn	25	55	1	26	1,83								
Fe	26	54		27	1,93	56		28	1,87	57	1	27	1,84

Tabelle 5.1:

For example, there are two stable isotopes ^{12}C of carbon with a content of 98.9% and ^{13}C with a content of 1.1%. This 1.1% must therefore contain such a Triteron core building block. Another example is oxygen, where ^{16}O (99.76%) dominates; the heavier isotopes are involved with only 0.037% (^{17}O) and 0.20% (^{18}O). ^{17}O is the isotope with the ^{3}He core building block. It is the rarest stable isotope of this element. Nitrogen is the last example: ^{14}N (> 99.5%) and ^{15}N (<0.4%). Again, it shows that the triteron core building block fits only in a few stable nitrogen isotopes.

The first element, where the triteron building block fits in all fluorine atoms, is the stable Flour-isotope ^{19}F. In contrast, ^{18}F converts with a half-life of 110 minutes in ^{18}O, which should actually consist of 9 stable deuteron building blocks, but has 2 triterons. Two triterons transform into three deuterons with the capture of one electron, and we get the 9 deuterons, but the L shell permanently leaves an electron. ^{22}Na, ^{26}Al, ^{30}P, ^{34}Cl, and ^{38}K behave similarly. ^{22}Na converts into ^{22}Ne with a half-life of 2.58 years.

They are all β^+ radiators with widely varying half-lives. The conversion may be by electron capture from the K shell, or by the exit of a positive charge in the form of a positron from the interior of the nucleus, which annuls with an electron from the environment to two γ-quanta. This raises the question of where the positron came from. Since the positron has exactly the same properties as the electron, but has a positive charge, it would have to split off from the proton. The difference between electron capture and positron emission would have to be noticeable in a mass difference of 2 electron masses in the nucleus, which would correspond to 0.001097 u. Unfortunately, the NIST data [5.39] of the radioactive isotopes are missing, so that this question can not be decided at the moment.

5.4.3 Table of the Radioactive Isotopes

The conversion of an element by electron capture is shown in Figure 5.3. A β^+-radiator incorporates an electron into the nucleus to stabilize the nucleus. In this case, its atomic number Z decreases by one while the mass number remains the same. The unstable isotope ^{10}C with 10 protons in 4 droplets, with two adjacent triterons producing a potential perturbation, capture an electron from the electron shell and each release a proton to the captured electron. The result is three deuterons. Figure 5.4 shows the spatial structure of the droplets of electrons and protons.

Edwin Kaal, who has presented a structured atom model (SAM) [5.36] on the EU2017 in Phoenix [5.40], has kindly given to me the magnets. The SAM is different from the present here droplet model because the magnetic balls represent the single protons while the electrons stays invisible.

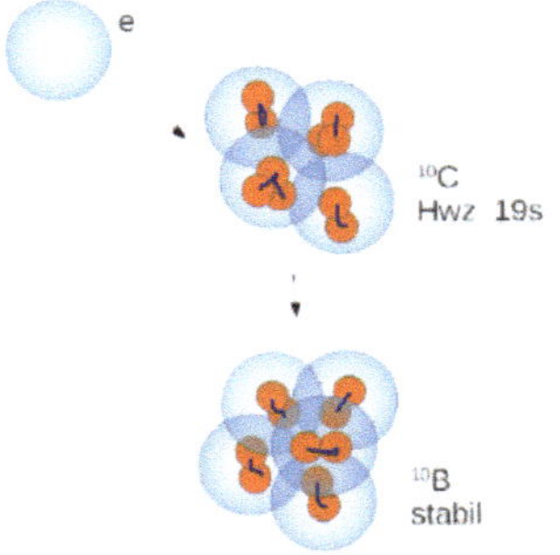

Figure 5.3: Electron capture

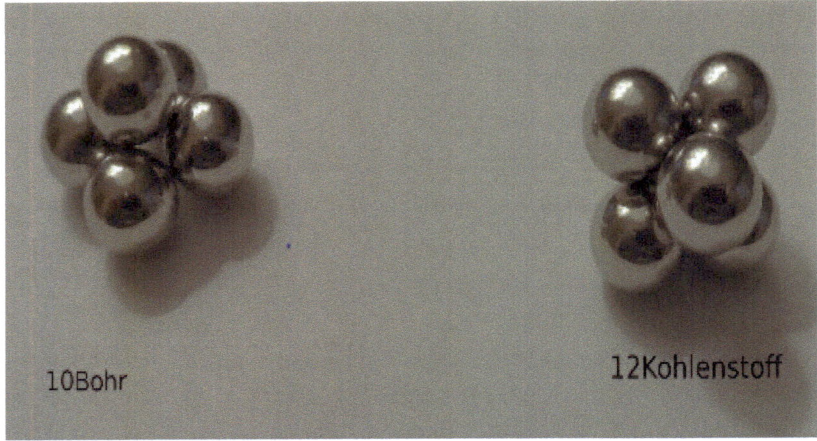

Figure 5.4: Core models modeled with magnets

The importance of the SAM model lies in the emphasis on the spatial structure of the arrangement of elementary magnets. In my droplet model, on the other hand, two protons in a nuclear electron form a droplet magnet. One single proton triplet in one droplet is allowed within the nucleus, hereinafter referred to as triteron. Two triterons decay with the capture of one electron into three deuterons, and if two droplets with only one proton (neutron) are in the nucleus, they combine to form a deuteron by giving up an electron. This makes it easier to detect the instabilities of isotopes.

Table 5.2 gives an example of a decay series with electron capture to a stable isotope. The gray line indicates the ratio of positive nuclear charge to negative nuclear charge, as well as the half-life.

	^{17}Ne (5T1D) + e^-	^{17}F (3T4D) + e^-	^{17}O (1T7D)
M/(M-Z) | Hlt	2,43 | 0,7s	2,13 | 66s	1,89 | stable
		^{21}Na (3T6D) + e^-	^{21}Ne (1T9D)
M/(M-Z) | Hlt		2,10 | 23s	1,91 | stable
	^{22}Mg (4T5D) + e^-	^{22}Na (2T8D) + e^-	^{22}Ne (11D)
M/(M-Z) | Hlt	2,20 | 3,9s	2,00 | 2,6a	1,83 | stable
	^{30}S (4T9D) + e^-	^{30}P (2T12D) + e^-	^{30}Si (15D)
M/(M-Z) | Hlt	2,14 | 1,4s	2,00 | 2,6m	1,88 | stable
	^{37}K (5T11D) + e^-	^{37}Ar (3T14D) + e^-	^{37}Cl (1T17D)

Tabelle 5.2

As can also be seen in the next table, the β^+-radiators all have a higher proportion of triterons. Each two adjacent triterons produce a positive partial potential disturbance and capture an electron from the electron shell. This gives rise to three deuterons and the potential against the shell is compensated again. Looking at the half-lives, these are very different, suggesting that when it comes to electron capture, the two causative triterons must be adjacent.

Beta-plus-Strahler

		Tröpfchen				Tröpfchen				Tröpfchen			
Element	**Z**	**M**	Tri	Duo	M/(M-Z)	**M**	Tri	Deu	M/(M-Z)	M	Tri	Deu	M/(M-Z)
Li	3									5	1	1	2,5
Be	4	6	2	0	3,00	7	2	0,5	2,33	8			2,00
B	5					8	2	1	2,67	9			2,25
C	6					10	2	2	2,50	11	1	4	2,20
N	7					12	2	3	2,40	13	1	5	2,17
O	8					14	2	4	2,33	15	1	6	2,14
F	9					17	3	4	2,13	18	2	6	2,00
Ne	10	17	5	1	2,43	18	2	6	2,25	19	1	8	2,11
Na	11	20	2	7	2,22	21	3	6	2,10	22	2	8	2,00
Mg	12					22	4	5	2,20	23	1	10	2,09
Al	13	24	2	9	2,18	25	3	8	2,08	26	2	10	2,00
Si	14					26	2	10	2,17	27	1	12	2,08
P	15	28	2	11	2,15	29	3	10	2,07	30	2	12	2,00
S	16					30	4	9	2,14	31	1	14	2,07
Cl	17	32	2	12	2,13	33	3	12	2,06	34	2	14	2,00
Ar	18					35	3	13	2,06	36	0	18	2,00
K	19					37	5	11	2,06	38	2	16	2,00
Ca	20					38	4	13	2,11	39	1	18	2,05
Sc	21	42	2	18	2,00	43	3	17	1,95	44	2	16	1,91
Ti	22	43	5	14	2,05	44	4	16	2,00	45	1	21	1,96
V	23	47	3	9	1,96	48	2	21	1,92	49	1	23	1,88
Cr	24	47	5	15	2,04	48	4	18	2,00	49	1	23	1,96
Mn	25	52	2	23	1,93	53	3	22	1,89	54	2	24	1,86
Fe	26	52	4	20	2,00	53	5	19	1,96	54	0	27	1,93

Table 5.3: β^+-radiators

Table 5.3 shows a few irregularities. ^{6}Be is an α-emitter. The two triterons do not turn into tree deuterons by electron capture, but decay into two α-particles which is obviously more energy efficient. ^{7}Be consists of two triterons and a neutron. Electron capture produces the stable ^{7}Li. ^{8}Be releases an α-particle and the stable ^{4}He is formed. ^{9}B releases an α-particle and the ^{5}Li is obtained, which in turn is an α-emitter and leaves hydrogen. Another exception is ^{36}Ar and ^{54}Fe. These are stable isotopes because they contain only deuterons.

Another category is the β$^-$-emitter. This emits electrons, so that the balance between the electron shell and atomic nucleus is compensated. The reason for this is a partial negative perturbation of the nuclear potential by two neighboring neutrons. The proton moves from one neutron to the other and transforms it into a deuteron. The vacated droplet is excreted as a free electron. When one electron is released, Z increases by one. The following examples (Figure5.5) show β^--decay series.

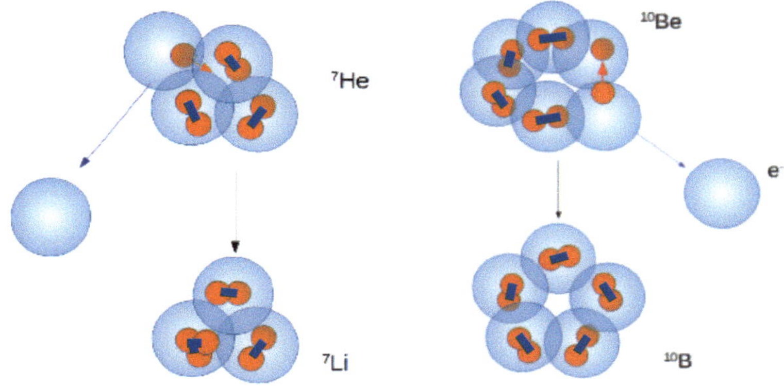

Figure 5.5: Electron donation

This conversion can only take place if the corresponding number of neutrons are contained in the nucleus. If there is only one neutron in the nucleus, the proton penetrates into a deuteron to form a triteron, as the left side of Figure 5.5 shows. However, when two neutrons are present, a proton migrates to a neutron and forms a deuteron, while the empty electron shell leaves the nucleus.

Table 5.5 shows two deviations for the Ca and Cr isotopes and one for the Fe isotope. There are provided some elements with black mass numbers in the right row M of Table 5.5. These elements are still stable, although the charge ratio is already well below 2. For the heaviest natural elements, such as lead and bismuth, the nuclear charge ratio can decrease to 1.65, and these elements still remain stable. But none of the elements that are heavier than bismuth remain stable. Here then is the border to the α-emitters, because by delivering an α-particle, can stabilize the charge ratio in heavy nuclei faster than by donating an electron.

	^{20}O (4n8D) – e$^-$	^{20}F (2n9D) – e$^-$	^{20}Ne (10D)
M/(M-Z) | Hlt	1,67 | 14,0s	1,82 | 11s	2,00 | stable
		^{23}Ne (1n11D)– e$^-$	^{23}Na (1T10D)
M/(M-Z) | Hlt		1,77 | 23s	1,92 | stable
	^{24}Ne (4n10D)– e$^-$	^{24}Na (2n11D)– e$^-$	^{24}Mg (12D)
M/(M-Z) | Hlt	1,71| 3,4m	1,85 | 15h	2,00 | stable
		^{31}Si (1n15D) – e$^-$	^{31}P(1T14D)
M/(M-Z) | Hlt		1,82 | 2,6h	1,94 | stable
	^{38}S (4n17D) – e$^-$	^{38}Cl (2n18D) – e$^-$	^{38}Ar (19D)
M/(M-Z) | Hwz	1,73 | 2,9h	1,81 | 37,3m	1,90 | stable

Table 5.4: Decay scheme

In terms of cosmology, however, only the elements up to the iron are of interest. Instead of assuming isolated nuclear forces, the atom was considered as one unit in this model. The atomic nucleus consists of units of neutrons, deuterons and triterons. Of these building blocks, only the deuteron is permanently stable and acts like an elementary magnet. A single triteron can be tolerated in a nucleus. Two triterons decay with the capture of one electron into three deuterons. Neutrons have the tendency to combine in pairs to form a pair of deuterons with the release of an electron.

A neutron and a deuteron transforms into a triteron with the release of an electron. A triteron and a neutron make two deuterons or an α-particle. This explains the decay rules. The transformations take place when these building blocks are adjacent to the atom and cause a partial potential disturbance.

Beta-minus-Strahler

		Tröpfchen				Tröpfchen				Tröpfchen			
Element	Z	M	n	Deu	M/(M-Z)	M	n	Deu	M/(M-Z)	M	n	Deu	M/(M-Z)
He	2	5			1,67	6	2	2	1,50	7	1	3	1,40
Li	3	8			1,60	9			1,50				
Be	4	10	2	4	1,67	11			1,57				
B	5	12	2	5	1,71	13	1	6	1,63				
C	6	14	2	6	1,75	15	1	7	1,67	16	4	6	1,60
N	7	16	2	7	1,78	17	1	8	1,70				
O	8	19	1	9	1,73	20	4	8	1,67				
F	9	20	2	9	1,82	21	1	10	1,75				
Ne	10	23	1	11	1,77	24	4	10	1,71				
Na	11	24	2	11	1,85	25	1	12	1,79	26	2	12	1,73
Mg	12	27	1	13	1,80	28	4	12	1,75				
Al	13	28	2	13	1,87	29	1	14	1,81	30	2	14	1,76
Si	14	31	1	15	1,82	32	4	14	1,78				
P	15	32	2	15	1,88	33	1	16	1,83	34	2	16	1,79
S	16	35	1	17	1,84	37	1	18	1,76	38	4	17	1,73
Cl	17	38	2	18	1,81	39	2	17	1,77	40	2	19	1,74
Ar	18	39	1	19	1,86	41	1	20	1,78	42	2	20	1,75
K	19	42	2	20	1,83	43	1	21	1,79	44	2	21	1,76
Ca	20	43	T	20	1,87	44		22	1,83	45	1	22	1,80
Sc	21	46	2	22	1,84	47	1	23	1,81	48	2	23	1,78
Ti	22	51	1	25	1,76								
V	23	52	2	25	1,79	53	1	26	1,77	54	2	26	1,74
Cr	24	53	T	25	1,83	54		27	1,80	55	1	27	1,77
Mn	25	56	2	27	1,81	57	1	28	1,78	58	2	28	1,76
Fe	26	58		29	1,81	59	1	29	1,79	60	2	29	1,76

Table 5.5: β^--radiators

Whenever this neighborhood occurs is unpredictable, but may be influenced by the atomic shell in an incomprehensible way. Statistically, this results in the half-life. Half-life is believed to be unchanging. However, with the decay of ^{187}Re in ^{187}Os realized in Darmstadt, it was found that strongly ionized atomic nuclei can decay faster than neutral ones [5.42]. Since this decay with a half-life of 42 billion years is used as a universal cosmic clock, this is problematic if the rhenium was ionized during the star development time. For the electron from the nuclear neutron it would be easier to leave the nucleus if the electron sheath were weaker. The half-life of ^{187}Re surprisingly dropped to 33 years with complete ionization. This should be similar for other elements and supports the proposed model.

The fact that the starting material of the nuclear fusion the hydrogen and the presence of deuterium on the earth only has a share of 0.3 ‰, and that in fused elements has a nuclear charge ratio of less than or equal to 2 for stable isotopes, has the consequence, that in space can not adopt a neutral environment, as standard cosmology asserts, but that stars, as sites of nuclear fusion, must be anodes and the dark depths of the cosmos must be sources of electrons and that the forces causing the movement are electromagnetic forces. As slightly more electrons than protons are consumed in the fusion of the elements to stabilize the nuclei, the surplus protons are blown into space as a solar wind. On the other hand, electrons can penetrate the sun and probably leave it again at the poles. The fusion products condense on the colder surface of the sun. We will discuss this hypothesis in detail later in 7.9 The Anode Model of the Sun.

The primordial nucleosynthesis discussed in the Standard Model, a few hundredths after the Big Bang assumes that with decreasing temperature in the cosmos the forming deuterium would not have been destroyed by high-energy photons and then a sequential nucleon synthesis would have taken place. This model, in contradiction to physics, assumes that the photon is not the result of a movement of charged particles. How then, does a chaotic movement, the temperature represent, suddenly produce such a web-like cosmic order as we observe?

The idea of an open asymmetric system without a big bang, far from the thermal equilibrium, seems much more plausible. If we remember that the world accepts itself as a self-similar system, then the deuterium described by Maxwell's equations with its sheath current and its core current is a transformer and one of the basic building blocks of the world. Maxwell's equations say nothing about the size of the circuits or where the current source must be. Like any ordinary transformer, the atom can receive its energy from outside and dissipate the unused energy as entropy. Below in section 6.6 About the Importance of Entropy in the Open System, we will take a closer look at the fundamental importance of this idea for understanding the world.

6 The Macrocosm

»Whenever a theory appears to you as the only possible one, take this as a sign that you have neither understood the theory nor the problem which it was intended to solve.« *Karl Popper*

6.1 A Review

We have seen in the last chapter how the approach affects the outcome of our observation. If we look at fast processes, we get a blur, just like when we extend the shutter speed of the camera. Nature is not blurred, but our perception is so sluggish, our engineering sense tells us. The Kantian "thing in itself" is not identical to our perception, we get at most a virtual image of the thing, but the positivist worldview clouds the mind because it makes subjective sensations the yardstick of the world. Looking at the macrocosm, there is the aspect of perspective from which we look at the world.

In a time before Nicolas Copernicus book *De revolutionibus orbium coelestium*, published in Nuremberg in 1543, one looked at the sky from a resting earth. The Ptolemaic world-view said that the stars were firmly attached to the centrals canopy, which is why they were called fixed stars. In contrast, the sun, the moon, and the visible planets orbited in circular orbits around the earth, although a few planets was observed with strange loop movements. This loops, Aristotle interpreted as a circular motion around a moving imaginary center superimposing the circular motion around the earth. The doctrine of Aristotle said that apart from the phases of the moon, there would be no changes in the sky, and that among the motions only the uniform circular motion would be perfect, and that it would proceed without external influence. The Copernicus' merit was that he placed the sun at the center of the world and thus introduced a heliocentric world-view, without, however, renouncing Aristotle's epicycles.

About 50 years after Copernicus's pioneering work, Johannes Kepler formulated the laws of planetary motion named after him, which were no longer based on epicycles, but on elliptical orbits, with the sun is in one focal point. Meanwhile, the Reformation wars had broken out in Germany and they dragged on Central Europe for more than a hundred years, which also weakened the power of the Catholic Church. Then Galileo Galilei succeeded in observing the Jupiter system with its moons and it became clear that Copernicus' hypothesis was confirmed and that the geocentric view of the world could no longer be sustained. The reaction of the church was correspondingly violent, its teachings resting on the geocentric world view of Ptolemy.

After another 100 years, Isaac Newton began to announce a new idea to the world. So far, the planetary movements were allegedly willed by God, thus he recognized a physical force behind the movement. Now he was able to explain ebb and flow and why an apple fell off the tree by mathematically describing how two masses attract each other. Likewise, he determined the escape speed $v = \sqrt{(2MG/r)}$ to 11.2 km/s, with a small body can leave the earth forever. To reach the circular path only one speed is required with a speed reduced by the factor $\sqrt{2}$. Our rockets are performed with these takeoff speeds to overcome the Earth's pull. But you need two mutually perpendicular forces for a circular orbit.

It took about another 200 years, when a young man named Albert Einstein, was searching for the missing forth, to explain a closed planetary orbit. It was clear that a body moves straightforward as long as there is no lateral force acting on it. However, for example, when a cyclist is moving on a sloped road in a curve, the perpendicular force of attraction and the resulting centrifugal force will cause a resultant steering force that keeps the driver on the path without his actively turning into the turn. Einstein transferred this naive idea to the conditions in the universe. The planets should somehow be held on invisible curved surfaces. Again, 100 years have passed since the publication of General Theory of Relativity, which has translated this idea to an obscure math.

With the onset of space travel in the middle of the last century and the further enhancement of our telescopes, we have opened the door to an interstellar and even an intergalactic viewpoint. If the progress of human knowledge continues to be so hindered by reactionary circles of the Church, I estimate that it takes another hundred years for an intergalactic point of view to prevail as general knowledge. The problem with general relativity isn't linguistic, but is related to the fact that Newton's law of force can explain why an apple falls on the earth, but not why the earth revolves around the sun. Einstein's proposed solution is movement on a curved surface that is sold not as a surface but as a space. It's the inflating balloon model with plenty of room for the Creationism around it.

A quote from Albert Einstein reads:

> *»Two things are infinite: the universe and human stupidity; and I'm not sure about the universe.«*

George Lemaître reconciling science with faith might have thought, if he is not sure, then we will make the universe closed and build the Creation around it. A bursting primal atom may well judge it and stupidity of the crowd will suffice to sell the whole thing to the faithful in as complicated a manner as possible.

Einstein's quote contains a sophistry. The word 'universe' includes the whole of the world. Consequently, it is infinite, since it contains everything known and unknown. On the other hand, completeness requires insularity. If he had used the term cosmos, it would have been clear that he meant the known part of the world, and he delimited the unknown part and kept the system open. Consequently, we will always speak here of the intergalactic cosmos, instead of the universe, meaning all light that our telescopes can capture today. We will have stronger telescopes in the future, so the cosmos stay open for us. May the incorrigible believers spread their Foreign-creation behind it, but not interfere into the discernible world.

Today's standard model of cosmology no longer corresponds to Lemaitre's idea of the primal atom in the details, but it has been preserved in the basic structure, the idea of an unimaginable primeval explosion [6.01]. As proof of this is considered by the supporters of the Big Bang the 1964 discovered cosmic microwave background radiation. It is a radiation whose maximum is 2.8 K and exactly corresponds to the average temperature of the cosmos, as required by the Planck radiation law for a black body. The measurements taken by the satellite mission COBE only confirmed that the cosmos has a temperature of 2.8 K from all directions and the found radiation anisotropy results from the movement of the solar system around the center of the Milky Way. On the other hand, the assumption cannot be substantiated that it is a remnant from the time of the Big Bang,

6.2 The Problem of the Missing Force in the Cosmos

It is said that Newton discovered his gravitational law from the quadratic decrease in gravity when an apple fell into his lap. In fact, his law is based on the exact observations of Tycho Brahe and Johannes Kepler, using the heliocentric world view of Copernicus. But when raindrops fall on an oblique glass plate, like on the roof of my sun-room, they do not run down the roof-light vertically, but describe an unpredictable way down, where they take more drops with that are in the vicinity. Consequently, not only gravity acts on a body, but there must be another force, which is indistinguishable from gravity, but acts per-

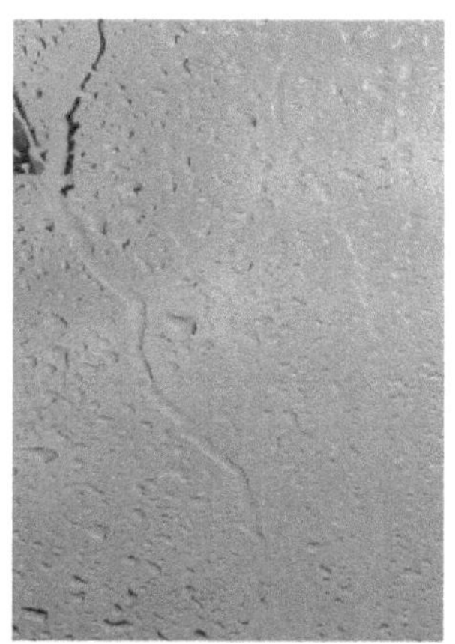

Figure 6.1: running raindrops on a roof-light

pendicular to it. And why do the fog droplets of the clouds stay up in the sky, but the raindrops of a certain size fall to the ground?

Any non-rectilinear motion is caused by two forces that act perpendicular to each other. The inertia of a body causes it to continue its rectilinear motion, as long as no force acts at an angle to the already existing force. Why should that apply not generally? Einstein wondered already. His idea, however, was a bit strange. He has undoubtedly heard in his Swiss mountains that you can drive through an inwardly inclined curve with almost no steering power. So a planet would have to move on an invisible inclined 'driving surface' within a closed curve. Only this idea does not work out. Not alone, the curvature does have to do with Newton's attraction between the sun and the earth but also with a force perpendicular to it, which would have to act evenly in parallel over the entire path diameter. When driving through a curve, no one expects a center of force at the pivot point of the curve. It only affects the slope of the road. But the sun is in the pivot of the earth's rotation. In that sense, the General Theory of Relativity can at best be regarded as a folly. He himself said to his friend Paul Ehrenfest:

> *»Ich habe schon wieder was verbrochen in der Gravitationstheorie, was mich ein wenig in Gefahr setzt, in einem Tollhaus interniert zu werden.« Translation: I've done something wrong again in the theory of gravity, which puts me a bit at risk of being interned in a madhouse.*

Nicely packaged in another foolishness, the reader has quickly lost track of when he has to fight through the scrub of Einstein's thoughts. There are already inferiority feelings in view of academic arrogance.

> *»The generalization of the theory of relativity was greatly facilitated by the shape given to Minkowski's special theory of relativity, which first made mathematicians clearly aware of the formal equality of spatial coordinates and the time coordinate and made them useful for theory.«* [6.02]

(As if formal equality of spatial coordinate and time coordinate! The space coordinate is a vector with one direction and time a scalar with no direction.) And below:

»The laws of physics must be such that they relate to arbitrarily moving frames of reference« [6.02]

Einstein wanted to overcome Newton's heliocentric viewpoint, what he explained very cumbersome and he was caught in the delusions of his relativity. If we take the observer for the reference frame, immediately becomes clear, that in the last quote, it doesn't have to say "t*hey relate to*" but rather "*they are independent of*". The mistake is in the condition. Physically, it is not the same thing, whether the earth is spinning around the sun or the sun is spinning around the earth, as he wants to make us believe. He would have had to take an intergalactic position to see the movement of the sun with its planets. If he had studied Dalton Miller's [6.03] experimental results more intensively, he might have come to other conclusions. Miller had found an anisotropy of the light with respect to the solar system's direction of motion [6.04], as later confirmed by the COBE data. Einstein has always denied these results and disqualified them as temperature differences.

»My opinion about Miller's experiments is the following. ... Should the positive result be confirmed, then the special theory of relativity and with it the general theory of relativity, in its current form, would be invalid. Experimentum summus judex. Only the equivalence of inertia and gravitation would remain, however, they would have to lead to a significantly different theory.« Albert Einstein, in a letter to Edwin E. Slosson, 8 July 1925

After Miller's death in 1941, Shankland, a disciple of Miller, after intensive consultations with Einstein, subjected Miller's work to criticism in the manner of Einstein and subsequently largely destroyed the data, as James DeMeo reported [6.03]. However, compared to Fresnel's entrainment effect of a mass flow, Miller's data are much too large and Einstein's theory of relativity is not confirmed by the negative declared Michelson-Morley experiment, since observer and light source are in the same frame of reference. I will come back again on Miller's experiment in the next chapter.

Back to our raindrops: It can be assumed here that they are slightly electrostatically charged and that the water molecule's dipole property means that the raindrops attract the glass plate. Now, the electrostatically attractive forces between the raindrops to gravity are comparatively small, since they only become visible when the glass pane has an inclination of approximately 20 degrees. Then the gravitational force affects only about one-third on the raindrops that drip off the glass plate. But what if the drops of water are smaller and more electrostatically charged? Everyone has experienced a thunderstorm before and knows about the power of a lightning discharge. Now we have two forces, the Newtonian and the Colombian force. The difference between the two is on the order of 39 powers of ten. Both Newton's Law and Coulomb's Law describe a force acting between two stationary bodies and weakening with the square of the distance. Both forces can be measured with a torsion balance. The moon would have long ago fallen like the apple to the earth, if there was not another force that we have not described. In order to drive a circular arc, every car driver has to turn his steering wheel so that he is not carried out of the curve. Each circular motion in the plane, like the motion of raindrops on the glass, is described by two forces that are perpendicular to each other, as illustrated in Figure 6.1. What force should that be? Anti-gravity? No, that would be just opposite to the gravitation, so it cancel the force directed to the center. It must be a force perpendicular to the force of attraction. The resulting force of both leads one body around the other on its orbit.

It is one of the biggest inconsistencies in physics to divide forces into electrical force and gravitational force, as well as weak and strong nuclear force. You measure the force in Newton meters and can only distinguish them according to their effective direction. Forces work where masses are always and constantly. They have no delay in their effect. Why then this strange distinction? There is an argument. Electromagnetic forces can be shielded by free electrons by breaking the magnetic field lines in the conductor. So there must be free electrons, as follows from the considerations in the previous chapter. Gravity can not be shielded because it follows from the dipole behavior of the atomic nucleus and atomic shell, as Wallac Thornhill has explained [6.05]. There-for gravity can not be modulated too. Consequently, no waves can arise on the gravitation. It turn about an electrical residual force that results from the charge transfer between the shell and core within the atoms as they aggregate. Such an arrangement is not a resonant circuit, such as the transition from an energetically higher to an energetically lower state, where the energy difference can be radiated.

We have learned that force is the product of mass and acceleration. Everyone has an idea of a mass independent of the object shape. We should remember here the Plasticine from our childhood, to the acceleration we experience when we start our car. It is the change of velocity due to a force. But what is mass? We remember, we determine masses by comparison with a standardized original kilogram. However that still does not explain mass. Everything that we know about the mass, we owe to mass spectrometry. It gives us the mass of the smallest building blocks of matter due to the fact that all masses carry a positive or negative electrical charge, which in the normal case almost cancel each other out, but are separated to differing degrees in the magnetic field according to their mass. The fact that we can measure the electric current and the magnetic field in the mass spectrometer allows the conclusion to the exact mass as a quantity of current at different positions of the detector.

So we need the knowledge of electrical engineering to answer the question of the second force for a curved movement. There it is the Lorentz force, which is perpendicular on the Coulomb force as in the mass spectrometer. For today's astrophysics, there is no special equation that describes this vertical force. Astrophysicists try to explain this difficulty with the erroneous curvature of the space of general relativity. But there are no curved spaces, only curved surfaces. Masses have a volume that you measure in a space spanned by three independent directions. Consequently, forces physically can only be distinguished by their strength and the three independent directions. If the astrophysicists were engineers, they might have noticed that according to their equations, the moon had fallen off. So the theory of relativity only obscures the problem and lead in a dead end. We have to turn back and rethink everything. Obviously, this is hindered by huge political power structures that determine scientific opinion, as was once the case in Galileo's time.

It remains to be clarified why this mischief has been able to survive for so long and is today accepted doctrine. The key to an explanation can be found in the work of Father George Lemaître, who barely made any publicity in the scientific public, whose work for a reconciliation of religion and science in the background was very effective. With the creation of a system for the evaluation of all scientific essays by peer groups, effective censorship was created in the sense of the Catholic Church, without its part being visible, because the peer leaders themselves are integrated into the academic world.

6.3 Lemaître's Idea of the Compatibility of the Creation with Science

»The Bible teaches us the path to go to heaven, not how the heaven is made.« *Galileo Galilee 1616*

From 1923 we find the newly ordained priest and doctor of mathematics Georges Lemaître for a year at the University of Cambridge, where Arthur Eddington introduced him to Stellar astronomy and numerical analysis. It was Eddington who was so firmly convinced of the correctness of Einstein's theory of relativity that he organized the solar eclipse expedition, which he organized together with astronomer Frank Watson Dyson, on the volcanic island of Principe in the Gulf of Guinea in West Africa on May 29, 1919 led to spectacular success. Although it was surprising to see, how under the worst possible conditions in the tropical midday heat amateur equipment yielding measurements of an accuracy that is today currently available only with the Hubble telescope outside the atmosphere [6.06].

From an early age Lemaître wanted to become a priest and a scientist. As a 17-year-old, he moved from a Jesuit school to the Catholic University of Leuven.[6]) The guest role in Cambridge was due to the destruction of the University of Leuven during the First World War. In Leuven, he began to write down his ideas for the expansion of the universe. The base of his ideas can already be found in the ancient Vedas of Indian mythology, the knowledge of which should not be unusual for a budding priest.

> *»This universe only existed in the Divine Idea, still unexpanded, as if it were hidden in darkness, imperceptible, indefinable, , not yet revealed, as if it had been lost in deep sleep.*
> *But then the self-existing power appeared in full glory, not perceptible, but which makes the world perceptible, with five elements and other principles of nature, expanding its idea, dispelling the darkness.*
> *He, who is perceived only with the inner sense, whose essence vanishes from the outer senses, who has no visible parts, that exists in eternity, even he, the soul of all being, appeared in form.« out of "The ordinances of menu", Sir William Jones from 1794* [6.08]

For the first time his work appeared in 1927 in the *Annales de la Société scientifique de Bruxelles* [6.09] a rather little known journal. [6.10] A translation into English titled *The Primeval Atom: An Essay on Cosmogony* was published by Van Nostrand in 1950 in New York, of which I made a translation into German. After scatterbrained speculating over the space, he developed the idea of a primeval giant-atom falling victim to radioactive decay and thus initiated a phase of expansion according to "known mechanisms," by which he meant the spectral redshift. [6.11] He gave Einstein his work for evaluation. Whose judgment was:

---

6 The Jesuits all over the world today run colleges, schools and boarding schools, providing general educational content to more than two million young people. Jesuits serve both Catholic and secular schools. The Order maintains 114 universities worldwide, where it significantly shapes education in philosophy, mathematics, astronomy and physics. In the twentieth century, the Jesuits defended the encyclical "Pascendi Dominici gregis" by Pope Pius X. [6.07])

»Your calculations are correct, but your understanding of physics is disgusting.« [6.12]

Einstein did not want to give up his position with a static space. Hubble's discovery of the spectral redshift, was the safe proof of a dynamic universe for Lemaitre, although Hubble was against it. Obviously, the spread of communism increasingly threatened the Christian world view, did the Catholic Church seek a savior, similar to the communist saviors Marx, Engels and Lenin to save the Occident for themselves?

According to the encyclical *Pascendi Dominici gregis* against the modernism of Pope Pius X, it can be assumed that the whole Catholic Church, and with it Lemaître too, was concerned that in Western Europe more and more people were falling away from the Christian faith. Unlike Pope Pius X, however, a number of Vatican priests sought reconciliation with science without, however, giving up their claim to leadership. Therefore, a new world view had to be created which had both a creed and a pseudo-scientific claim, so as to make science serviceable to theology again. In this idea the priests might have been agreed in the circles of the Catholic Church of this time. Since St. Peter in the 1st century, the Roman church had the sovereignty over the kingdom of heaven. Galileo had already scratched at it and Darwin, this traitor, had replaced the doctrine of creation with the doctrine of evolution. Should now also be lost the authority over the sky? Something had to happen. Because Einstein's paper *On Electrodynamics of Moving Bodies* was based on the divine order and symmetry, and this World basis was still generalized until 1915, thus creating the egg of the Vedas, it was only necessary to speed up and inflate it so that it would take the form of a chalice. This was done by Alexander Friedman [6.13] in 1922 independently of Lemaitre, who solved Einstein's equations and found a singularity that could be interpreted as expansion or mass concentration, depending on the choice of parameters. Hubble's research came just in time. So the inflation of the universe was justified to the spectral redshift, as it were interpreted as speed after Doppler. Now, in contrast to the relativity, the inflation process only had to be set to the beginning of an absolute time and it was derived from the reciprocal of the Hubble constant and obtained the value of 13.58 billion years. The world of God then lay before the act of creation and behind the radius of the universe from the product of the speed of light and the age of the cosmos. This restored the harmony between science and religion ...

Einstein was not alone in judging Lemaître's ideas, that they were totally unacceptable. This was the opinion of almost all scientists too, but Lemaître managed to persuade them[6.14]. Einstein rejected a cosmological constant as a stabilizing factor in 1931 and is said to have described it as the greatest folly of his life.

The National Socialists had suspended 1933 a head premium on Einstein. They persecuted him as Jews. He met Lemaître in 1933 in Brussels and could not return to Germany because of the Nazis. At this time, he was met with the invitation to relocate to Princeton by Abraham Flexner, who had founded a Model Center for Higher Education there as early as 1930, the Institute for Advanced Study, Princeton, New Jersey. As the first director of the institute (1930-1939), Flexner gathered some of the most important scientists of his time from around the world. In 1921 Einstein had received the Nobel Prize and then he had been to Princeton for a lecture. There Lemaître visited him again in 1935 and he must have managed to change Einstein's mind. In any case, through his alliance with the Catholic Church, he became an icon of science, but his career came to an end because, despite his popularity, he was increasingly isolated scientifically, as his letters to Velikovsky showed [6.15]. He refused to quantum mechanics and also he rejected black holes.

Einsteins view of the geocentric and the heliocentric world in his general relativity, has, become the keystone of catholic religion. Thus, the harmony between faith and science established by Lemaître was accepted by the Pontifical Academy of Sciences, and in recognition of his services to the doctrines of the Church, Lemaître received his appointment to the Pontifical Academy of Sciences. In 1951, the Pope Pius XII declared, that Lemaître's theory of the Big Bang was the scientific validation of Catholicism. He stated in a lecture, that the beginning of the world, which can be defined in time with the Big Bang, originated in a divine act of creation. Lemaitre received a number of church honors, so followed in 1960 his appointment by Pope John XXIII as President of the Pontifical Academy and the honorary title Monsignor. Lemaître and Daniel O'Connell, a Jesuit and from 1952 the Pope's scientific adviser, convinced the pope not to mention creationism publicly anymore [6.16]. That was not necessary after all.

The work of the Jesuit educational institutions is due to the spread of Lemaître's ideas. Thus, this theory was anchored not only in science, but also in religion and has become the official teaching program of all Catholic and secular universities. If we look around, we find that the number of Catholic universities has grown strongly internationally, and therefore there is not only a spiritual, but also a material interest in the preservation of this theory [6.17]. People are told that the true nature of the physical world can not be understood, except for a few geniuses, such as Einstein and Hawking, who would be able to think in four dimensions. Physics is something you have to believe, not what you can understand. It is the religion for those who no longer believe in the Immaculate Conception of Mother Mary. The physical and sexual misuse of subordinates by criminal priests, much discussed in public today, always begins with the mental misuse, the suppression of the truth, or the creation of one's own, a Catholic truth, for the sake of maintaining power. Obviously, a common commitment to fiction strengthens the cohesion of social groups. The content of this fiction is obviously of secondary importance. Thus, such fictions can take effect beyond religious and national borders too, even though they contradict common sense. Perhaps theorists have recognized this nowadays and use this fiction to secure funds for new and ever more expensive experiments as long as society allows it.

»Science without religion is lame, religion without science is blind« - Albert Einstein

The original quote is from Immanuel Kant and reads

»Thoughts without intuitions are empty, intuitions without concepts are blind. ... All human knowledge begins with intuitions, proceeds from thence to concepts, and ends with ideas.«

6.3 Black Holes

As early as 1783, John Michell a Briton, speculating on dark stars, believed that gravity would be strong enough to trap light. In a letter published by the Royal Society, he wrote

»If the radius of a sphere of the same density as the sun exceeds that of the sun in a ratio of 500 to 1, a body falling upon it from an infinite height would have a higher velocity on its surface than that of light. Consequently, assuming that light is attracted to the same force as other bodies by its mass, any light emitted by such a body would return to it as a result of its own gravitation. This is true on the assumption that light is affected by gravity in the same way as massive objects.« [6.18]

The idea that corpuscular light could not escape from 'heavy' stars was also described by Laplace in his Exposition du Système du Monde in 1796. He created the term 'dark body' (corps obscure) [6.19]. The term *black hole* was introduced in 1967 by John A. Wheeler. In the outer space of sufficiently compact masses, a space region characterized by an event horizon is to form, into which matter can only fall in but can not come out again. This should also apply to mass-less electromagnetic waves, such as visible light. So was his faith. However, this definition was brightened over time as people began searching for suitable objects in the cosmos. Eventually that caused quite a bit of confusion, as the following quote shows.

»Black holes are not made of matter, even though they have a big mass, which explains why it has not been possible to observe them directly but only through the effect of their gravity on the environment, distorting space and time and having an irresistible force of attraction. It's hard to believe that the idea behind such exotic objects is more than 230 years old.« [6.20]

The quote also shows how little people understand of physics who talk about black holes. It is especially embarrassing because the quote is on a website of the German Max Planck Society of 2017. If black holes are not material, they are obviously ideally. How then they can have big mass? How do one want to prove that the observed effects are not optical, but are caused by gravity, which we can not measure in the cosmos over a great distance? How do one want to fix a space curvature when curvature of light is an optical effect at phase transitions and how is a black hole to be observed at all if it does not allow light to pass to the observer? According to Stephen Hawking, however, it should give off thermal radiation [6.21], as he would have calculated. There is no way to verify the accuracy of the statement, unless you look for a black hole and convince yourself of it. However, black holes have not yet been reported in the systematic mapping of the sky from the **S**loan **D**igital **S**ky **S**urvey project (SDSS) [6.22]. As well as? Observation requires light. There are the black spots, which are retouched in photographs of galaxies, nothing but deliberate stupefaction of citizens. As of 2007, I personally classified over 4,700 images of galaxies from the SDSS database as part of the 'Galaxy Zoo' research project according to Hubble's catalog of shapes [6.23]. There was no black hole in any galaxy. On the contrary, all galaxy centers were characterized by a stronger light in the center than in the more distant regions.

As I sit over the manuscript, a press release actually comes with a picture of a ring-shaped structure and it is claimed that the interior of the ring is a black hole. It is in the ring structure to radio radiation [6.24]. In addition one must know that the radio radiation originates from molecules. In the optical spectrum, the center of a galaxy is very bright, indicating that the molecules there are split into atoms and ions, so that no radio radiation can come from the center of a galaxy. So it's a new scam in science.

Karl Schwarzschild found in the trenches of the Eastern Front in World War 1 the solution of Einstein's equation of gravitation. [6.25] It actually consists of two solutions, The outer solution has the shape of a funnel and the inner solution is a hollow sphere. Under the influence of an increasing mass, the funnel he believed narrows and opens at the bottom. whereby all masses including light are sucked into an ever-narrowing funnel. That's his black hole. The masses carry with him no electrical charges. However, since the mass from the hole can not disappear, the mass should get an infinite density in a zero volume, which is physically impossible. The paper went unnoticed for a long time, until Stephen Hawking, the dark lord of the black holes, continued to spin the yarn for "Pope's New Clothes," making a curious invisible fabric out of it.

Figure 6.2 shows the rotation surface of the Schwarzschild solution and embedded in it a star with a circulating planet. The first question is to ask for the observer. Where is he sitting, is he sitting in the funnel or on the star or at the planet? A look at the Schwarzschild equations says he's sitting in the funnel. The space coordinate ct should be the distance between the observer and the light source, but neither *c* nor *t* is a vector, also light has no preferred direction. The direction of view is also the axis of rotation of the funnel.

The principle of relativity says that all points of view should give the same picture because physics would not change. The axis of rotation of the funnel would therefore move with the observer and the observer would never come out of the funnel. We see, however, from the outside on the funnel. Physics does not change, but perspective, and positivism not does demand to describe the reality, but the image I observe. Consequently, from the positivistic point of view I must be able to interchange star and planet , or observer and star. Anyone who has held a camera in his hands knows that he get a different picture from every point of view. On the planet you get the planeto-centric view of Aristotle, If you shift the observation point to the star, you get the view of Galileo. If you take an interstellar point of view, you don't get a picture, which Schwarzschild shows us.

Figure 6.2: Schwarzschild solution with real Sun (yellow)and Earth(blue)

Clever people could argue that way: That's just a three-dimensional model. In a four-dimensional reality, that looks very different. If the space has four dimensions, then the planet and the star also have four dimensions, and a four-dimensional hyper-sphere then rolls on a three-dimensional hyper-surface and the ratios would be the same, as we see in Figure 6.2. We had already stated in section 2.5.1 What is a Space? that this fourth dimension is a fake, since it determines the direction from the light source to the observer, as the metric is defined. The center of gravitation is not the center of the light source, because then the funnel would degenerate to a plane, which Newton presupposed. We see that the equations give a completely different picture than what is preached to us by the high priests of physics. Dos that fatally remember to the time before Luther's Reformation, doesn't it, where people could not read the Bible because they didn't understand Latin? Today we have the situation that people can not read the mathematical formulas and so there is an unprecedented freedom of interpretation in the sense of maintaining the power of inherited structures of religion and the for it serving research.

If black holes are not material, the star can not be at the bottom of the funnel, but it must be in the center of the planetary orbit, otherwise the beam direction of the light would be superfluous. The gravitational force must then be uniform and parallel to the axis of rotation on the entire surface of rotation Schwarzschild solution acting from below from the yellow surrounding funnel in figure 6.2. Then this corresponds to the conditions of a cornering on the earth's surface. According to the principle of relativity we would have to be able to interchange star and planet and model relations should not change. The different masses of both bodies are unlikely to affect gravitation, as it is caused by the curvature of the space and has nothing to do with the masses of the bodies. In other words, the sun could turn around the earth again. However, only the case was observed that the larger bodies were in the center of the movement. The generalization of the image into the four-dimensional case turns the simple rotation surface into a hyper-surface and the three-dimensional space into a four-dimensional one. A generalization is an operation concerning the change of quantities, not the change of quality. As a result, the surface property is retained, as is the space around it. But time has no direction and the speed of light also does not have, so there is no independent 4th space direction. That's what the equations do say! At least now, you should be convinced of the absurdity of the theory of relativity. The audience has been pretended something completely different. This is the technique of magicians who understand the art of distracting the audience's attention. Here attention is directed to the solution of the equations and the imagination is turned off. "You must calculate and imagine nothing," we were told as students. That is idle in this book to sift through all the corners of this composed physics building of modern physics. If you want to do it anyway, Paul Marmet's book *Einstein's Theory of Relativity versus Classical Mechanics* [6.26] is very recommended.

We will always come up against the intention to strengthen the ancient faith that the Church has been representing for over two thousand years. With the theory of relativity, it has been possible partially to restore the geocentric world view. This is the stated mission of the Jesuits, who have taken on the teaching duties there, where the state has withdrawn. The methods have only become more subtle. The Immaculate Conception of Mother Mary has had its day. Today, they are attacking people in their biggest weakness, mathematics. It is grotesque to observe that mathematics, on the one hand, has evolved into a tool of engineers that has produced the computer and fantastic equipment for exploring cosmos, and on the other hand the massive using of computers on many people simultaneously shut down the sagacity.

6.5 The Battle against Black Holes

Academic questions are strange and often unworldly and their answer is often even curious: *How many angels fit on a needle point?* was one of a curious question that had been discussed in the founding period of the University of Jena. For fun, I've googled the question. The result was about 20 hits, which dealt directly with the question. There I read, that was an assumption of Christian Morgenstern, who had attacked the scholasticism in a poem. It would not have been angels, but souls and the correct answer would be 1000.

An equally curious scholastic question is this: if something falls into a black hole, is it then lost forever? A quarrel about this question, arose between Stephen Hawking and Leonard Susskind, which dragged on for many years. The answer is no less curious. We learn about it in the book *The Black Hole War, My Battle with Stephen Hawking to Make the World Save for Quantum Mechanics* [6.27]. Hawking argued in 1983 that if something falls into a black hole, the information can not be reconstructed. Hawking obviously did not mean information, but messages [7]). Susskind argued that everything he knew about quantum the-

7 Shannon distinguishes the message from the information. The signal from which the message is generated by coding is sent through the information channel, decoded and forwarded to the recipient as a message. The information channel has a certain transmission capacity for this. That is a quantity. The concept of information is thus a completely syntactical term without semantic content. In contrast, there is another interpretation of information in quantum mechanics. http://philsci-archive.pitt.edu/10911/1/What_is_Shannon_Information.pdf

ory was destroyed by this assertion. For him, black holes are a reservoir of entropy. He speaks of hidden information, obviously he means hidden message not information. Hawking had already brightened the black holes bit in the search for suitable candidates in nature, by announcing that they would give off thermal radiation and they could also evaporate over time, as I mentioned earlier. The holographic principle (which Susskind co-developed) would say that the third dimension is an illusion and that energy and matter are only forms of information. If they fall into a black hole, then the conservation of matter would no longer be possible. According to Susskind, at the 1993 Santa Barbara Conference, Hawking claimed that his theory said that his black hole radiation contained no information, that it would be purely thermal (So entropy is dissipated!). The thermal noise in the transmission channel is according to Shannon information. (At first, it has to be clarified whether a black hole is an information channel in the sense of Shannon.) Only Susskind, t'Hooft and a small number of participants disagreed and trusted the connection between entropy and information. Susskind reports the following about Hawking's lesson:

> *»For Stephen, the CGHS [8]) meant that the mathematics of theory simply proved his point of view. For me, not only was the mental point of view wrong, but also the mathematical justification of quantum gravity, which is ultimately embedded in the CGHS, contradictory.«* [6.26 p.250]

Finally, there was a vote. Four options for voting were proposed and the vote produced the following result

1. 25 votes for Hawking's option, the information is lost.
2. 39 votes for T'Hooft's and Susskind's option. The information is preserved and comes back with Hawking radiation.
3. 7 votes for the information is caught in Planck length-large remnants.

---

8 CGHS named after Callan-Giddings-Harvey-Strominger two-dimensional model of general relativity with a space dimension and a time dimension

4. 6 votes for other.

This vote was still rather undecided. In his above-mentioned book, Susskind propagated a no less curious string theory, which also proved to be a dead end, as Lee Smolin has pointed out in his book *Trouble with Physics* [6.27].

The interesting thing about Susskind's book mentioned above is not the technical aspects, but the thinking and behavioral patterns of a small group of people who behave like a sect, are arrogant and want to determine the development of a whole discipline. So we read in [6.25 p.251]:

> *»What about Stephen, who attracts the attention that a holy man can receive, who wants to reveal the deepest mysteries of God and the universe?" Hawking is an arrogant man, very of self-absorbed, extremely self-centered, and the same for half of the people I know, including myself, and I think the answer to that question is partly the magic and mystery of the disembodied intellect driving the universe in its wheelchair But part of that is that theoretical physics is a small world made up of people who have nothing other know for years.«*

Hawking died in 2018, his brain was not disembodied. It only didn't work in the parts responsible for the muscle work for a long time and this inevitably also the intellect harm because it also strong reduces the sensory abilities. After 30 years of tough wrestling, Hawking was beaten. From his reasoning why he gave up, here's a quote:

> *»However, the situation changed when I discovered that quantum effects cause a black hole to emit radiation,«*

Hawking continued, referring to his 1974 findings:.

> *»From the approximations I used, this radiation would have to be completely thermal and therefore could not carry any information. So what would happen to all the information trapped inside the black hole when it evaporates and eventually dissolves completely? It seems that the only way information can come out of it is that the radiation is not exactly thermal but has subtle correlations.«* [6.28]

It is something wonderful with 'pure' reason. You feel so close to God. Only that deductively you can not gain new insights, has not got around in theoreticians. Stephen Hawking, for all his fame, has not contributed anything to new insights, as Hilton Ratcliffe has noted:

> *»No matter how hard I am, I can not find anything really useful in Dr. Stephen Hawking entire span of his career what I would classify as a good science. One of the best spirits currently on the planet gave its best shot over half a century ago, became world famous, acquired tremendous wealth, was worshiped and defended by millions, is recognized and respected in the farthest corners of the world, and for what? It's a shocking realization, but based on what he has produced in his life, Dr. Hawking is one of the most ineffective scientists as long as anyone remember.«* [6.29]

Closed systems tend to die off. This applies not only in physics, but also in society, as we had bitterly to recognize at the end of the GDR. In this respect scholasticism is not a method that promises lasting success because it closes itself to the new. Even if there are 500 years between the first and the second question, from the point of view of systems theory they belong to the same category and they lead to similar answers. There are neither immaterial souls nor black holes in the real world. What we mean by souls is the information of the structured material nervous system. Therefore, it is better to put his knowledge on paper than to pray to God for salvation. Because the messages for posterity are not the information variables that are understood from a thermodynamic point of view, the bits can each contain two values, but their concrete messages, which are encoded therein.

Static mass-bound information, arises at the phase boundary of two matter forms, but this has nothing to do with the holographic principle. 'Living' information is transmitted by the energy of a vibrating medium. It must be realized that the electric charge and its movement is the cause of all forces. Then it becomes clear that there is no external force that can put matter into the state, that it loses its inner structure and thus any information. There certainly will be Followers of the theory of black holes in future too, only black holes will never be measured. Consequently, they are uninteresting to physics. That the third dimension is an illusion should dawn on the dreamy Susskind at the latest when he has run against a lamppost. In string theory he still calculated with 9 dimensions. So it does not matter to him and his colleagues to explore the real world at all. They only do their business in order to collect a lot of research funds and in order to achieve that, they are always talking about God. My recommendation:

Do not take these people seriously when they throw around cryptic symbols and say they calculated something without having measured anything before.

Here is a quote from St. Hawking on theories:

> *»A physical theory is only a mathematical model to describe the results of our observations. A theory is a good theory if it is an elegant model if it is a large class of observational objects Moreover, when it predicts the results of new observations, it makes no sense to ask if it is true because we do not know what reality is independent of a theory.«* [6.30]

Who determines what the results of the observations are? The theorist? And he does determine what the reality has to be, does he? This is presumptuous in the extreme. As Schopenhauer said: "*The world is my will and my imagination*!" We must credit Stephen Hawking for revoking his theory in January 2014 with the article *Information Preservation and Weather Forecasting for black holes*. He says there:

> *»I take this as indicating that ... there can be no event horizons and no firewalls.* ***The absence of event horizons means that there are no black holes*** *- in the sense of the regime, from which light can not escape to infinity.«* [6.32]

But if there are no black holes, a gravitational wave triggered by two black holes is a phantom. Norbert Lossau, the chief editor-of science of the 'Welt', wrote on Feb. 16th 2017, three years after Hawking revoked his black holes,

> *»Black holes merge more often than previously thought. The LIGO researchers have observed a whole series of such fusions. Who will receive the Nobel Prize for these insights? «.* [6.33]

It's hard to recapture a stupidity that was once circulated, and the Nobel Prize was actually awarded for the 'Gravity Wave Detection' in 2017. Nobel had decreed that the prize for humanity's most useful research accomplishments be awarded every year. Perhaps someone can tell me conclusively what the benefit to humanity is of looking after a phantom in the face of global environmental degradation. No, the stupidity is topped by presenting the public now an alleged photo of a black hole from the center of M87. Of course, a spectroscopist does not expect molecular radiation there, because molecules in this area break because of the high radiation density in the visible range.

One more remark about the predictive power of a theory.

You can never draw more knowledge from a theory than you put into it.

At this point, I would like to recall the failed attempts to build a perpetual motion machine, that should do the work without being energized. That these attempts failed, the second law of thermodynamics is responsible for the increase in entropy in closed systems. A completed theory is a closed system. More knowledge means more order too. But new insights come only from new sensory experiences from the outside. So we will consider now open systems.

6.6 About the Importance of Entropy in the Open System

To equate the entropy with information is obviously just as wrong as equating the information channel with the information. Entropy is therefore to be understood as an information carrier, just like the mass or energy, but with the difference that this energy or mass is no longer usable for the system. So it is rubbish in the traditional sense. For its part, energy is bound to a carrier too, to mass. And mass generates the force field for the transmission of information. In a closed system, entropy increases as the usable energy decreases. This is what the second law of thermodynamics says.

$dS_{int} / dt \geq 0$

It is the phrase that has repeatedly brought physics into severe distress by breaking the symmetry and now it turns out, that it is also incomplete, since it applies only to closed systems. In physics, closed systems only exist as idealized models. Real systems, like any atom, are thermodynamically open: they exchange at least radiation energy with the environment, larger systems even exchange mass. But the second law of thermodynamics does not say anything about that. Comparing this law to Indian Vedanta philosophy, it embodies the principle Shiva.

But what does the principle of creation and the principle Vishnu of preservation look like? The Indian Gods are seen as a real part of nature, comparable to Western pantheism. Personalization that is to be understood as avatars serve rather as representatives of these principles. So the idea of a foreign creation is unknown there and Charles Darwin with his evolution theory is not so far from the Indian principle of the divine Trinity. Today we talk about self-organization. Ilja Prigogine expanded equilibrium thermodynamics to thermodynamics of open systems far from thermodynamic equilibrium and was awarded the Nobel Prize in 1977.

$$dS_{system} = dS_{inp} - dS_{out} + dS_{int} = dS_{ext} + dS_{int} \qquad (6.01)$$

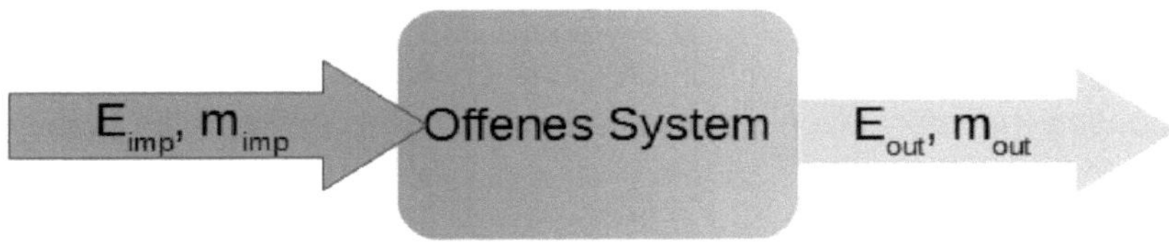

Figure 6.3: Open system with exchange

In an open system, there is no equilibrium, but a stationary state in which the brought in energy is equal to the energy released and the brought in mass is equal to the delivered mass.

What can we say about the entropy in such a system? The entropy change of an open system is the sum of the entered entropy change and the internal entropy change minus the emitted entropy. We summarize the difference between registered entropy and emitted entropy as external entropy. Then we can write:

Let us look at the first case of equation (6.01), the case of creation. If you create something, it is equivalent to creating a certain order. This means that you have to clear out.. (It does not mean that something arises from nothing. If you multiply something by zero, the result is always zero. Gods also have to bow to the law.) So the system entropy is total negative. $dS_{system} < 0$. For this to be the case, the external entropy must be negative and more entropy must be removed than is left internally and is added. You can express this through the relations:

Creation: $dS_{ext} < 0, \quad |dS_{ext}| > dS_{int}$

The order remains as long as the system entropy does not change. $dS_{system} = 0$ This is expressed mathematically by the following relations:

Conservation: $dS_{ext} < 0, \quad |dS_{ext}| = dS_{int}$

A system can perish due to internal and external factors. This happens when the entropy of the entire system grows. The classic 2nd law of thermodynamics applies here. $dS_{system} > 0$ for

Destruction:

1. Decay of the order of **internal factors**: less entropy is dissipated by the system than is generated internally. To describe

this case, we use the relations:

$dS_{ext} < 0$ $|dS_{ext}| < dSint$

2. The decay of order through **external factors**: the external entropy is greater than the internal. This expresses the following relation:

   $dS_{ext} > dS_{int}$

It can be seen from these few relations that a moderate use of natural resources is the first divine commandment and not the growth of profits at all costs. One of the four noble truths that greed only creates suffering was recognized more than 2000 years ago in the Orient only up to the Occident this truth has not yet penetrated. It wouldn't does matter if the suffering only hit the greedy. Their suffering is that they can never satisfy their greed and so the divine nature has to suffer with.

Let's look at our Earth: Our Earth is in a radiation balance. This means that the external entropy must be negative. In other words, the entropy imported from the sun must be smaller than the entropy exported from the earth.

The imported entropy is $dS_{inp} = Q/T_{sun}$. The heat quantity of the sun results from the product of the solar constant times the illuminated area times the time. Since only half the area is illuminated, Q results in $S{\cdot}\pi{\cdot}r^2{\cdot}t$. With S = 1360 W/m², r = 6,371 × 10^6 m and T_{sun} = 5800 K, we obtain Q = 1,7×10^{17} J for 1 second

dSi_{np} = 1,7×10^{17} J/5800K = 3,3×10^{13} J/K

The release of energy to the surrounding space takes place at a temperature of about 260 K. This is the temperature of the earth's surface that a high altitude observer would measure. This results in the entropy export of the earth: $dS_{exp} = Q/T_{earth}$

dS_{out} = 1,7×10^{17} J/260K = 65×10^{13} J/K

$dS_{ext} = dSi_{np} - dS_{out}$ = (3,3 – 65) × 10^{13} J/K → $dS_{ext} < 0$

$|dS_{ext}| = 61{,}7 \times 10^{13}$ J/K $\quad > \quad dS_{int} = 0{,}5 \times 10^{13}$J/K

The value of dS_{int} is calculated from the world energy consumption of 2008, which is converted into heat at least 2/3 according to the Carnot process

The earth is a self-organizing system.

Wherever structures form, we are apparently dealing with open systems whose external entropy is negative. However, the equations also show that the system can tilt, if the internal entropy becomes larger than the external one. This is the case when more energy and mass is spent than resources are available. Obviously it is not enough just to look at the incident radiation and output radiation of the heat at the moment.

The divine Trinity, or better said, we can now mathematically formulate the basic law of evolution. When Einstein read the holy book of the Hindus, the Bhagavad Gita, is said he to have uttered.

> *»When I read the Bhagavad Gita, I wonder how God created the universe, everything else seems superfluous.«*

as I learned in a hotel in Rajasthan, when I read the back cover of an Indian edition of this sacred book lying on the bedside table instead of the usual Bible in Europe. That has not let go of me over the years and I wanted to know, how that could be meant. Here my question has now found an answer, an answer with which, in my opinion, a believer as well an atheist can live.

At this point I will come back to Maxwell's equations. The equations describe an electrical and a magnetic vortex, as we saw in 5.4, so we can now consider an atom as an open system. Then it receives energy as $div\ E = \rho/\varepsilon$ says and emits entropy in the form of electromagnetic radiation. An atom acts like a Tesla transformer. It has an outer current loop in the atomic shell and an inner current loop in the core. The protons form the magnetic core. It is not the union of quantum theory and relativity that is the key to understanding the world, but the relationship between energy and entropy at the phase boundaries of the bipolar matter. We were only unable to decipher the ancient knowledge of the divine trinity of creation, preservation and destruction because we had lost the meaning of the words.

With this answer in mind, we can now put aside the old relativistic world view and embark on a new, a world view based on the idea of open systems, flooded with electrical energy, without a known initial state.

»God is nowhere treated worse than by naturalists who believe in him.«
Friedrich Engels

7 Outlook in an Intergalactic World

»Everything we know comes from the light.« - *Jim Ryder*

Nowadays we have telescopes that we use to collect light from the furthest corners of the cosmos to admire God's lamp shop. We have built interstellar probes that left our solar system and brought close-up shots of the planets that reveal phenomena that are incomprehensible with our inherited ideas of the past century. We can now take an interstellar viewpoint and see that our Sun moves in the Milky Way around a center whose properties we want to learn to understand because this movement does not correspond to the ideas that neither Newton nor Einstein have developed. If the sun is not fixed in space, then the planetary orbits are not ellipses but helix orbits and if Newton's law is no longer valid within the galaxy, that does not mean that we have to invent gravity monsters and dark matter to explain these phenomena. All we have to do is remember what we see, and that is the light that conjures structures to our night sky that carries all the information that needs to be decoded. We can measure light according to intensity and spectral wavelength.

We can not yet measure the forces acting in the cosmos. Let's keep track of what we can observe and measure. That should be the basis of our future considerations. For this we need basic knowledge in systems theory, thermodynamics, electrodynamics, spectroscopy and plasma physics. But do not worry, that sounds more demanding than it is. Above all, we need common sense. These are not classic disciplines of astrophysics. They have their roots in the engineering sciences and have developed there also in concrete applications. The space technology of the last 60 years did the rest to get a completely new picture of the cosmos.

But what use is it to us? That we can leave our earth and move on when we have destroyed our planet is an illusion that no one with clear mind can believe in. The benefit could be to learn to understand the energy sources of the cosmos and to reproduce them on Earth on a much smaller scale in order to be able to replace the fossil energy sources. Instead of starting from a belief that is the standard model of cosmology, we should trust our technical experience and understand the cosmos as an expansion of our horizon, in which everything works according to the same rules as on Earth, only on a much larger one Scale. It is not important to know whether there was a beginning and what it looked like. We'll never figure that out. But it is important to recognize that our sun and its planetary system are not isolated in the cosmos. It moves as a mass point in a much larger system and from there it draws its energy and releases the rest as radiation entropy, a tiny part of which hits the earth.

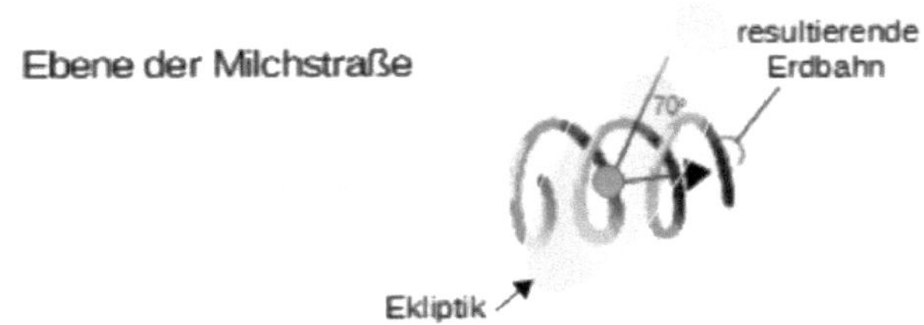

Figure 7.1 The planetary helix from the intergalactic view, schematic representation

If the sun rotates once in about 250 million years at an average distance of 26,500 light years around the center of our galaxy, the Milky Way, is that 6.3 billion kilometers per year. Then the sun, in turn, with its planetary system, moves at its own speed with circa 230 km/s [7.01] according to NASA around the center of the Milky Way, and one day we realize that the sun orbit will be no simply a circular orbit but the sun performs rotationally or oscillatory movements perpendicular to its orbital plane.

We recall that G. Miller found from the anisotropy of light a value of 208 km/s for the earth drift together with the planetary system on Mount Wilson with his interferometer (Figure 7.2) in 1933. This is only 9.4 % less than the value, deduced from a Galactic Year according to the Russian astronomer Pavel Petrovich Parenago from 1952. He found 225 million Earth years for a solar orbit around the galactic center, and with the distance of the Sun from the center of the Milky Way, this gives 222 km/s.[9]). This is an astonishing concordance, which was only possible with a highest care with which the measurements were carried out under the conditions of the time.

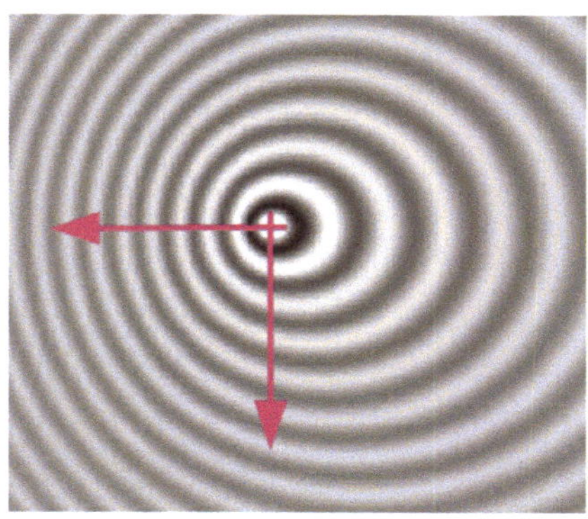

Figure7.2: Anisotropy of a moving light source, the interferometer measures the difference in the number of rings over a fixed distance

In June 1921 Einstein wrote to the physicist Robert Millikan:

> *»I believe that I have really found the relationship between gravitation and electricity, assuming that the Miller experiments are based on a fundamental error. Otherwise, the whole relativity theory collapses like a house of cards.«*

9 Distance to the galactic center: 26,500 light-years; --> $2\pi \cdot r$ = 166504.41 light-years = 1 575 298 223 010 000 000 km circumference
Galactic Year = 225 000 000 Earth Years = 7 096 500 000 000 000 s -->
Speed = circumference / Galactic Year = 221.98 km/s

as reported by DeMeo. [7.02]. In the Cleveland Plain Dealer newspaper of Jan. 27, 1926, Miller commented as follows:

> *»The trouble with Prof. Einstein is that he knows nothing about my results. He has been saying for thirty years that the interferometer experiments in Cleveland showed negative results. We never said they gave negative results, and they did not in fact give negative results . He ought to give me credit for knowing that temperature differences would affect the results . He wrote to me in November suggesting this . I am not so simple as to make no allowance for temperature.«*

To support the big bang theory and thus the theory of relativity, Miller's results had to be obliterate. This did not succeed completely in the 20th century. However, for years, his research had been forgotten. This scandal is at least as great as the condemnation of Galileo by the Roman Catholic Church in the 17th century, except that this time she did not cause a stir and made work others for themselves.

At the beginning of the 20th century, one still had the idea of an entrainment effect of an ether wind. But the explanation for Miller's discovery is different. We know the Doppler effect of light, which, when a light source moves, causes an anisotropy between the direction of motion and perpendicular to it, which manifests itself in the displacement of the spectral lines. This anisotropy obviously changes with altitude above the Earth's field. On Mount Wilson, the earth's field is perpendicular to the observation direction of the interferometer and thus has no influence on the measurement result. In contrast, the gravity field is disturbed directly on the ground by surrounding buildings and land elevations. Thus, the mass of the galaxy at height forms a new physical reference system. Miller was the discoverer of this effect and one should not withhold this honor for him.

The helix orbit angle of the Earth in the galactic frame is 81.5 degrees. This high pitch angle of the earth's orbit is due to the fact that the earth's rotation is about six and a half times slower than the forward motion of the sun. We remember the lacking force that made Einstein ponder already. But to get a helix requires three forces, all of which must be perpendicular to each other, because the principle of inertia says:

A body flies straight out as long as there no external force acting on it.

Newton has identified the centripetal force as a gravitational force, but there are still two forces left to be identified to explain the motion of our solar system along the Milky Way. Modern astrophysics has no answer to that.

7.1 The Pioneers of the New Worldview

Einstein's fateful belief that the relativity theory could eliminate electrodynamics from the cosmos was his conviction until the end of his life. The discussion with Velikovski [7.03] did not change that, who triggered a worldwide scandal with his book *Worlds in Collision* [7.04] by questioning the laws of celestial mechanics. This scandal was not without consequences for all involved in the publication of the book. That was in the middle of the 20th century.

At the beginning of the twentieth century, at a time when Einstein was still considering the theory of relativity in the Zurich Patent Office, the Norwegian physicist Kristian Birkeland [7.05] returned from his polar expedition and, risking his life, brought home the realization that electric currents from the Sun are the cause of the nocturnal illumination of the sky in the polar regions. He reproduced the results in the laboratory and actually received the observed effect of an Aurora borealis even under laboratory conditions. This pioneering act was the birth of a new cosmology. Meanwhile, he received his due recognition by being depicted on the Norwegian 200-crown bill. In other countries, however, he is largely unknown in public.

The second pioneer of this new cosmology was Hannes Alfvén [7.06], [7.07], a Swedish electrical engineer who continued Birkeland's work in 1939 and studied terrestrial magnetic storms. He gained new insights in the field of magnetohydrodynamics and recognized as early as 1963 the fibrous structure of the cosmos. He received the Nobel Prize in Physics in 1970 for "his fundamental findings and discoveries in magnetohydrodynamics with fruitful applications in various parts of plasma physics" but his conclusions for the cosmology and publications on this topic remained largely unnoticed. The specialization in the physical disciplines had already advanced too much and the theoreticians had become obsessed too much with the theory of relativity. Measurements on the magnetosphere and the electrically charged solar wind, as well as the investigations of the Pioneer Voyager missions on Saturn brought Alfvén a completely new view of the cosmos in the eighties [7.08]. The probes confirmed Birkeland's discoveries of the electric currents between the sun and planets.

The cosmos isn't relativistic but electrodynamic!

Everything that shines and rotates in the cosmos is in the plasma state and therefore has electrodynamic causes and no relativistic, as Einstein claimed, when he "took a seat" on a fast electron and thus the electrical forces noticed any more. We only get direct knowledge of the stars from the electromagnetic spectrum, not from the forces that there govern, including Newtonian force, which for a century occupied the minds of astrophysicists. But to get an idea of the cosmic movements, it is important to identify more forces based on existing structures. A glimpse into the history of the discovery of the cosmic motion sequences should remind us that it was Copernicus who discovered that the Earth revolves around the Sun and that with Hubble's realization that our solar system embedded in the Milky Way has its own motion, the Movement of the planets is not on closed paths, but describe helix paths along the solar orbit. In the course of the twentieth century, the realization grew that the cosmic structure is networked, and that the galaxies are suspended in this network, like drops of dew in a spider's web.

A third pioneer must be mentioned here. Antony Peratt published the formation laws for spiral galaxies in 1986 by simulating a plasma stream on a super computer [7.09]. Eric Lerner summed up the knowledge of plasma cosmology collected until 1992 under the title *The Big Bang Never Happened* [7.10] and explained how a quasar works.

7.2 The Birkeland Current

Our experience tells us that the lamps are powered by electricity in God's lamp shop. If we introduce God as the operator, we need not worry about the origin of the electricity, but we also know that according to section 6.5, the lamps are consumers in an open self-organizing irreversible system. God is the wild card for our temporary lack of knowledge. At least we avoided the big big bang. According to Kirchhoff, the network of currents can be described by the mesh and node rule and Ohm's law. Here, however, it is interesting to see how a plasma stream in the cosmos looks like.

Maxwell's laws are not just applicable to the propagation of light. They also describe the four basic principles of electromagnetism.

1. A changing magnetic field generates an electric field perpendicular to the direction of change of the magnetic field.
2. A changing electric field generates a magnetic field perpendicular to the direction of change of the electric field.
3. An isolated electric charge causes the attraction of the opposite charge.
4. there are no isolated magnetic poles.

The luminous cosmic threads consist of plasma filaments between the network nodes on which the galaxies are strung like pearls or dewdrops. Such a plasma filament is characterized by the Birkeland current, a plasma helix. as shown in Figure 7.3. This plasma helix is named after the Norwegian plasma pioneer Kristian Birkeland. We see three forces on this helix, the ***Coulomb force*** responsible for the current flow, the ***Lorentz force*** acting perpendicularly to the tangent to the magnetic lines, and one third force, which in turn is perpendicular to the Lorentz force. It results from the pinch effect and should therefore be called Pinch force. The **Pinch force** results from the effect of the Lorentz force, which causes a twisting of the two plasma filaments. You can best understand on a rubber engine, used in aircraft modeling, how forces work in a Birkeland current. Imagine our planetary system embedded in a Birkeland current, Newton's force acts exactly in the same direction as the pinch force, and apart from mass with its charge, there is nothing that could cause this force. The perpendicular Lorentz force ensures that the planets do not fall into the sun. Thus, we can inductively conclude that the stars and planets are dust grains in these streams and are carried along by these three perpendicular forces. From that follows: Newton's gravitational force is in any case be attributed in part to the Pinch force in a Birkeland current.

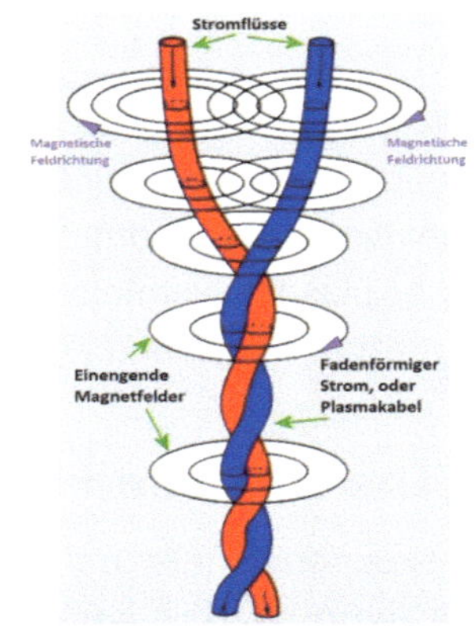

Figure 7.3 Birkeland current
Reference: Thunderbolts.info

At first glance, it is amazing that gravity should be the cause of the interaction of the magnetic fields of the planets with the magnetic field of the sun. On the other hand, it is evident that an undisturbed Birkeland current creates a double star system that rotates around a common center of gravity in the stream. If we look at the Birkeland current on average, as shown in Figure 7.4, we see that the Lorentz force is perpendicular to the Coulomb force, which causes the flow of current and acts tangentially to the magnetic field lines.

The pinch force ***F***, which pulls the two currents together, is again perpendicular to the Lorentz force, which eventually together cause the twisting of the two parallel current conductors. If a small body moves in this magnetic field, it describes an elliptical helix around the two currents, which are located in the two focal points of this ellipse. In the event that one current dominates the other, only one current will remain and the ellipse will only rotate around the one focal point of the ellipse. The result is the well-known perihelion movement that we observe by Mercury. We must not forget that the sun is a positively charged body [7.11], which moves itself, which of course generates a magnetic field, and that the moving planets also have or have had magnetic fields, when they were in the plasma state in past. The calculation of potential flows is the subject of Differential Geometry and Partial Differential Equations [7.12] and moreover of magnetohydrodynamical fluid mechanics [7.13]. Here to enter deeper into, is beyond the scope of this book, especially the used mathematics goes over the high school level. Only the equation of a circular helix is given here for the sake of completeness which shows the three directions:

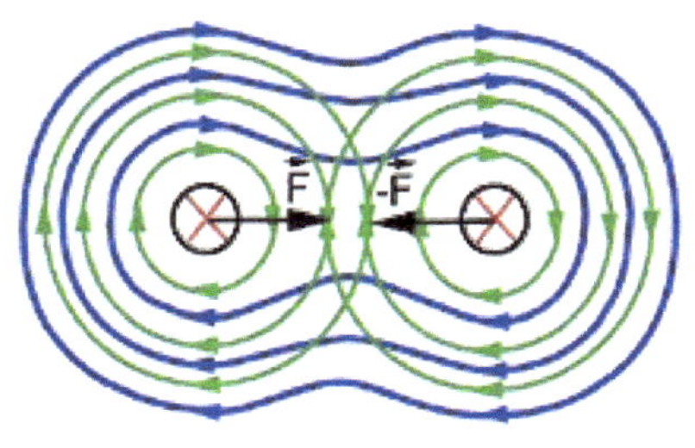

Figure 7.4: Cross section through the Birkeland stream

$$\boldsymbol{r}(k) = a\cdot(\mathbf{e}_1\cdot\cos k + \mathbf{e}_2\cdot\sin k) + b\cdot k\cdot\mathbf{e}_3 \qquad (7.01)$$

This equation is a vector equation with $\boldsymbol{r}$ as the vector describing the helix determined by the unit vectors $\mathbf{e}_1$ to $\mathbf{e}_3$ in R^3. k is a variable that can take all values between *a* and *b*. *a* is the radius of the helix and $2\pi b$ is the pitch of the helix.

Thus, the rather complex planetary motion comes about through the three mutually perpendicular forces described by Coulomb, Newton, and Lorentz. If you want to understand the motion of the celestial bodies, you have to consider them in the galaxy as an open current-carrying system. On the other hand, the galaxy itself is an open system in the large Birkeland current that connects the cosmic network nodes.

The structure of the Birkeland stream can be modeled mathematically. The Swedish physicist Stig Lundquist, a student of Hannes Alfvén, proposed a model of a force-free magnetic field as early as 1950 [7.14]. This work was only rediscovered by Don Scott in 2015 [7.15]. From Maxwell's equations, the equation follows for cylinder coordinates

$$r^2 \frac{\partial^2 B_z(r)}{\partial r^2} + r \frac{\partial B_z(r)}{\partial r} + \alpha^2 r^2 B_z(r) = 0. \qquad (7.02)$$

Lundquist examined whether magnetic fields could exist in an electrically conductive liquid and presented, among other things, the now known Bessel solution for *force-free fields*.(It means weightless) It is now generally assumed that magnetic fields in interplanetary magnetic clouds and plasma currents in the solar photosphere are force-free. According to the Bessel function, these fields do not have a continuously decreasing field strength, but a periodically changing field strength, which shows the typical image of the iron filings around a current-carrying conductor. The iron filings are aligned according to the Bessel function in the magnetic field as Figure 5.1 has shown.

If the current is to be force-free, this means that the Lorentz forces cancel each other out at different distances from the center. This results in a self-organized characteristic structure, by which a Birkeland current can be recognized.

7.3 Cosmic Velocities

The determination of the velocity of celestial bodies in the cosmos becomes more and more difficult the farther they are from the earth, since a change of position is no longer perceived in manageable times. We depend on an indirect method, the displacement of the spectral lines due to the Doppler effect. But this effect has proved to be less reliable, as Halton Arp, a student of Hubble, has noted in meticulous observations [7.16]. This has something to do with the different density of particles in the cosmos. The allegedly empty space is indeed traversed by a network of dark streams of charged particles, which we call plasma, and which, due to the infinite widths of the cosmos, despite its extremely low density, could make up a large part of the mass in the known cosmos. The heavenly bodies swim in these streams. Magnetic fields separate the charges and the particles strive for recombination, which we perceive as glow between the celestial bodies.

Due to the different masses of protons and electrons, the two particle types have very different velocities, which can not be treated within classical thermodynamics, because the electron temperature differs significantly from the ion temperature. Here comes a property to effect as seen in a stream of particles of different velocity. The faster and lighter particles overtake the heavier slow particles. When overtaking, they have to dodge, which leads to a change of direction. Now you might think that this overtaking would take place quite disorderly. That would be quite a mess. In fact, one observes a very orderly passing process, as happens on our motorways. Thus, a self-organized vortex forms under the influence of a magnetic field surrounding the electrically charged particles. However, recombination requires an approximation of both velocities below the rate of ionization. To be able to recombine, the electrons must be decelerated. Thus, depending on the density of the plasma and thus on the free path length of a charge carrier, there are different behaviors of the plasma. If the plasma is very thin, it is in dark mode. if it becomes increasingly denser, it comes to the glow mode and at high densities it comes to arc or lightning discharge. This, in turn, presupposes that before there can be a mass discharge, there must have been a charge separation that have built up over a longer period of time. Due to the extremely high velocity of the electrons compared to the protons, which are 1836 times heavier than the electrons, the electrons are not immediately recaptured by the ions. It comes to plasma oscillations. However, it has been observed that at velocities of more than 160 km/s, these plasma oscillations disappear [7.17],[7.20]. The formation of a Birkeland current is therefore a resonance phenomenon of the electrons, which forces the ions into a current. There is therefore a lower limit of the velocity v_{pu}, where the plasma starts to form the typical properties of the Birkeland current.

In flowing plasma there then three characteristic velocities are for the protons. The second characteristic velocity results from the maximum kinetic energy a proton can absorb in relation to an electron by ionization of a Hydrogen atom:

$$v_{po} = \sqrt{\frac{m_e}{m_p}} \cdot c \qquad (7.03)$$

The upper velocity v_{po} is 7000 km/s caused by the impulse of light speed of electron and the third is the geometric mean of the lower plasma velocity v_{pu} and the velocity according to equation (7.03):

$$v = \sqrt{v_{pu} \cdot v_{po}} \qquad (7.04)$$

That gives 1058 km/s. Calculating the escape velocity of an electron for the ionization of hydrogen gives a speed of 1026 km/s. If one encounters luminous hydrogen clouds, you can assume that the difference in velocity between electrons and protons is on the order of a little over 1000 km/s, because the speeds between electrons and protons for recombination must approach so far as. This speed around 1050 km/s is therefore the mean velocity of glowing hydrogen plasma. Larger luminous atoms have a lower velocity because of their mass.

1050 km/s, that is the speed with which protons in plasma threads are traveling in the cosmos.
Larger luminous mass particles are correspondingly slower.

The velocities of cosmic objects are also limited according to their mass by the speed of light. Solving equation (4.16) for *v*, you get the upper bound for velocity:

$$v = c \cdot \sqrt{1 - \frac{E_0^{\,2}}{E^2}} \tag{7.05}$$

Only charged particles, which are accelerated in very strong magnetic fields, reach speeds that are approximately comparable to the speed of light. The solar wind has an upper speed of only about 750 km/s, as NASA reported [7.16].

7.3.1 Can the Spectral Redshift be Interpreted as Speed?

In 1916, the first 2.5 m reflector telescope at the Mount Wilson Observatory in Pasadena was completed and an employee was sought. This employee was Edwin Hubble. However, as a result of the entry of the United States in the 1st World War, he could not begin his work until 1919 at the Mount Wilson Observatory. He was particularly interested in the nebulae that had not yet been recognized as galaxies at that time. This insight was reserved for Hubble. Another finding was that the brightness of certain variable giant stars in these nebulae varied in such a way that their period of fluctuation was directly related to their absolute brightness. By comparing this absolute brightness to the observed apparent brightness, Hubble was able to determine the distance of the star in a distant galaxy, and thus the distance of the galaxy itself. He found that the distance of many galaxies depended linearly on their spectral redshift. The proportionality factor was later called the Hubble constant. He published these results in 1929. However, throughout his life he refused to acknowledge the redshift z as a measure of galaxy escaping speed, and with good reason, as we shall see.

First, we ask: **Can the redshift of spectral lines of galaxies be interpreted as their distance?** The redshift z of a spectral line results from the line spectrum:

$$z = \frac{\lambda_w - \lambda}{\lambda} \tag{7.06}$$

where λ_w is the wavelength observed on a spectral line of a star or galaxy spectrum and λ is the wavelength measured on the earth. Here the Hubble law is to serve for the estimation of the distance of a galaxy. Distances can be determined exactly only by stereometry. As a result, exact measurements are obtained only within 100 astronomical units (three light-years = 1 parsec). (See also section 3.4.4 Limits of Observability) Greater distances are based on always more less reliable assumptions that are not verifiable. This also applies to the Hubble correlation, because the absolute brightness of a galaxy can not be measured. It is based on the assumption that the cosmos is absolutely translucent and that the galaxies are all about the same size, which can be safely accepted in certain areas. But this is no longer valid in the vicinity of neighboring galaxies. In addition, images of the associations of two galaxies have already been obtained. According to Hubble, the distance *r* results as:

$$r = \frac{z \cdot c}{H} \; Mpc \tag{7.07}$$

Where *r* is the distance in megaparsec, *c* is the speed of light in the cosmic medium, and *H* is the Hubble parameter[10]), which is the slope of a regression line measured between the redshift and the absolute

10 The Hubble parameter was corrected again and again. It has been reduced by a factor of 7 since Hubble's time. The assumed value is about 70 km/s·Mpc. This shows how unreliable all distance information based on redshifts is.

brightness of the galaxies. With $R = c / H = 4286$, we obtain a constant with which only z is to be multiplied, and we obtain the distance in Mpc, but z should be smaller than 0.4. We have learned in section 3.7 Modeling, Experimental Design and Measurement that physical equations are valid only in the measurement range and require a clear demarcation or bound. This barrier of 0.4 for z follows from the correlation of the spectral redshift with the absolute brightness of the galaxy up to a deviation of 10%. For larger values of z, the deviation quickly becomes very large, which means that we no longer receive reliable distance information. In other words, the absolute light transmission is no longer given to these distances. Further, we must note that the equation applies only to free-standing galaxies. Halton Arp has found a number of cosmic objects near other objects, for which the equation is no longer valid, since the cosmic medium is then denser there than in the rest of space.

In section 7.3 we estimated the upper limit velocity with which we can calculate the maximum expected wavelength shift due to the Doppler effect of galaxies. According to our calculation example for the sound from section 3.6.1 we can now compile the data for the calculation. We want to calculate the maximum Doppler shift for the H_{α}-line of hydrogen, since this line can be found in all optical spectra. It lies in the red region at 656.3 nm. This results in a frequency of 4.571×10^{11} Hz. This results in a wavelength shift of 2.3 nm at an intrinsic velocity of the protons of 1058 km/s with respect to the observer on Earth. But because the line shift is not absolute, but relative, we have to divide by the wavelength according to equation (7.3) to get the size z.

$z = 0,0035$

If we multiply z with c we get the escape velocity. If we take the escape speed from (7.04) into account, then we see that z cannot significantly exceed an upper limit of 1. In 2008, however, a cosmic object with a redshift of 3.9 was reported [7.17], which was named APM08279 + 5225. At least since this finding the alarm bells would have to ring the bell for theorists because their theory collapses. But instead of thinking about it, they invented the 'alternative fact' with the expansion of space. Only they have not considered, if you stretch the scale similar to a rubber band, then the true distances seem to decrease. With the discovery of further extremely red-shifted galaxies EGSY8p7 with z = 8.86 and GN-z11 with z = 11.1 [7.18] the contradictions within the big bang model are further exacerbated.

This small amount is the proportion of the spectral redshift that can be caused by the escape velocity of galaxies. It is the amount that demarcates the Doppler effect from other causes of redshift. If this amount is exceeded, there must be another cause for the spectral redshift. Compared with the measured amounts of redshift in cosmic line spectra, the Doppler effect is a negligible factor, and the spectral redshifts suggest that the expansion of the entire cosmos is a huge inductive fallacy. Hubble was right that the spectral redshift can not be interpreted as escape velocity. H. Arp underlines this in his book about the intrinsic redshift [7.16].

7.3.2 The Energetic Causes of the Redshift

The Doppler effect of light is based solely on the fact that the light source and the observer move relative to each other. This movement does not affect the emission process of the light, because the relative motion of observer and emitting atoms that move all in the same direction has no influence on the emission process itself. In this respect, the shape of the emission line at rest does not differ from that in the movement. This is different when atoms move disorderly in different directions relative to the observer. The result is a Doppler broadening of the emission line. What this is about is the orderly movement of the light sources, which is superimposed on the disordered motion of the atoms. By plotting a relative line width function over the redshift z for a large set of galaxies we expect a spot field included in two straight lines that is parallel to the abscissa, one for the maximum and one for the minimum line width, because there is expected no dependence of the line width on the spectral redshift.

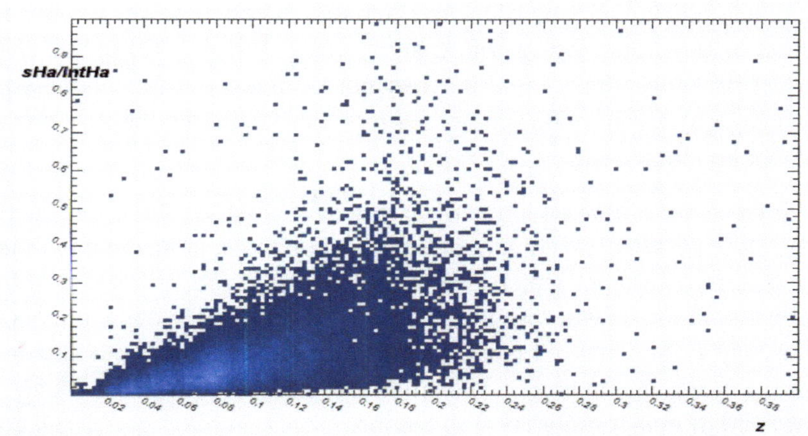

Figure 7.5: Relative line width of the H_α line depending on its spectra. Each point represents another active spiral galaxy [7.22]

Figure 7.5 shows a point field of active spiral galaxies where the relative line width of the H_α-line was correlated with its shift. There it turns out differently than expected that the two straight lines limiting the point field have a clear gradient. The conclusion can only be that between the galaxies there are processes, which simultaneously cause a line broadening to the redshift, which has to do with an energy decrease of the light over cosmic distances. This is called 'tired light'. These effects only appear beyond the barrier of z = 0.0035. An explanation for the 'tired light' can be found by Paul Marmet [7.21]. He found 109 quasars (quasi-stellar objects) for which both absorption and emission lines could be measured, and where the value of the absorption red shift was always different from the measured red shift in emission for the same object. It is clear that such results cannot be attributed to a Doppler red shift, which is interpreted as movement of the entire object. Careful investigation of the mechanism for the scattering of electromagnetic radiation by gaseous atoms and molecules shows that an electron is always accelerated for a short time as a consequence of the momentum transmitted by a photon (wave train). Such an acceleration of an electric charge generates Bremsstrahlung in the very long-wave range. As a result, a small amount of energy is lost to the photon, which manifests itself in a broadening of the spectral lines. It is the Compton effect, which is known for γ-rays, but can not be ruled out for longer-wavelength light, even though the scattered very long-wave photon has not yet been measured.

We have seen how an inductive conclusion, such as the application of the Doppler effect on light, needs to be bounded in its context, so that it does not contradict statements already recognized as true. It should prevail the realization that natural laws, unlike mathematical equations, not only in the units of measurement must be consistent, but also have areas of validity, even if we often do not know them. So we move forward from measured value to measured value in manageable steps to the limit. Astrophysics is particularly susceptible to the expansion of natural laws into uncontrollable areas. They will then fast establish a stack of hypothesis, without possessing any real basis of the experience more.

7.3.3 Quasars

Quasars were mentioned in the previous section, cosmic objects with puzzling spectral redshifts. Quasars are cosmic objects that have an extreme red shift in their spectra relative to galaxies and are therefore seen by standard cosmology at the edge of the known cosmos. The name Quasar derives from English. quasi-stellar object or quasi-stellar radio source. Although quasars were first discovered using radio telescopes, only about 10% of quasars actually send out radio emission. Therefore we differentiate between radio-silent and radio-loud quasars. Radio waves indicate molecular clouds, especially H_2 and CO but also other molecules.

We have stated above that *z* can not exceed the limit of 1, so as not to break the whole theoretical building of previous physics, if we want to interpret the spectral redshift as velocity. In our particular case, we have a redshift of 2.1, and that's not the upper limit for this class of cosmic objects. In December 2017, the journal Nature reported the quasar ULAS J134208.10 + 092838.61 with a redshift of 7.54 [7.23]

As a result of the presumed huge distance quasars were attributed an unusually high energy density. Where this energy density came from was a big mystery and so-called 'black holes' were thought to be the cause of the energy source, although they were defined as an energy sink and they were supposed to swallow all the radiation. In fact, the redshift has nothing to do with the removal of the quasars, as Paul Marmet explained in his article *The Cosmological Constant and the Redshift of Quasars.* [7.24] Remarkable are the broad spectral lines of multiply ionized carbon, silicon, magnesium, nitrogen and oxygen, as shown in Figure 7.6. The width of the emission lines indicates a high electron density of the plasma, which in turn indicates a high electromagnetic activity. By contrast, the absorption lines are comparatively very narrow.

Because a black hole is a theoretical assumption of an unrealistic theory, as we noted in the previous chapter, there must be another explanation for the luminance of a quasar. In research on nuclear fusion, the Russian Filipov and independently of him the American Mather, discovered a possibility to generate high-density plasma with little energy [7.27]. The generating device was named **plasma focus**. It consists of two concentrically arranged copper tubes, which act as electrodes. When a large current discharges across the copper electrodes, a donuts-shaped plasma node is created - a **plasmoid**. The term dates back to Winston H. Bostick in 1956. This plasmoid is exactly the opposite of a black hole. He devours no masses, but separates the charges of sucked masses by means of electromotive forces in two opposite directions. Eric Lerner, who met Bostick around 1974, came up with the idea that a quasar and a plasmoid, although very different in size, are identical in shape and dynamics. Both consist of an extremely dense source of energy emitting jets of high-frequency radiation in opposite directions like you see in figure 7.6 and 7.7. The plasma focus can increase the energy density to a factor of 10^{16} of incoming energy. Due to the high energy density, the atomic nuclei are fused and excited in the plasma and γ-rays are emitted.

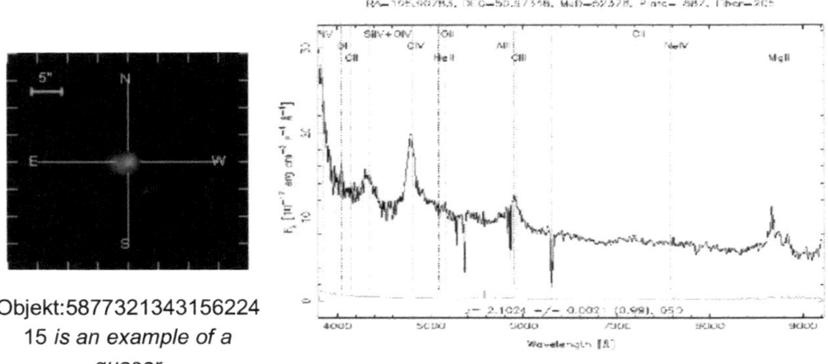

Objekt:587732134315622415 *is an example of a quasar*

The difference in line width between emission and absorption lines is clearly visible. Emission lines are hills an absorption lines are narrow canyons.
Reference: The Sloan Digital Sky Survey

Figure 7.6: Quasar with red shift z = 2,1
Reference: The Sloan Digital Sky Survey

What we now observe as quasars is one of the jets that is currently focused on the earth. When a galaxy rotates in a magnetic field, a current must be generated along the axis of rotation, as Hannes Alfvén [7.26] predicted.

Any plasma with enough energy will create vortex filaments and these filaments will grow without limits, if space and time allow it. While Lerner sees the galaxy itself as a quasar -

> *»A quasar is thus the birth cry of a galaxy, through which the excess rotational energy, which must be eliminated in the collapse of the proto-galaxy, is carried away in the form of energy beams.«* [7.27]

- due to the observations of Halton Arp you are concluding that a galaxy is able to produce several quasars from its center during its active phase. He summarized his observations in the Atlas of Peculiar Galaxies [7.28]. Especially the active Seyfert galaxies are candidates for the production of quasars from their center.

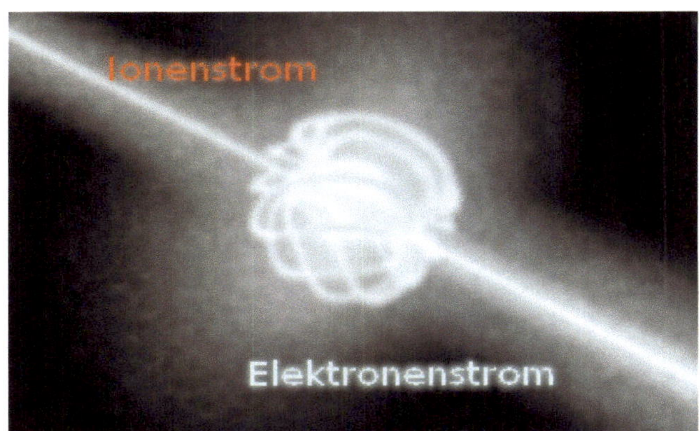

Figure 7.7: Plasmoid from a plasma focus by Lerner

Observations in different spectral regions have revealed filaments such as umbilical cords connecting a galaxy to the quasar. Over its live time, galaxies produce multiple quasars that line up like a string of pearls, slowly moving away from the center of the galaxy's axis of rotation. Their redshift decreases with age.

To learn more about quasars, the databases of the Sloan Digital Sky Survey Project (SDSS), a project for the systematic capture of all cosmic objects, can be analyzed. The SDSS catalog [7.29] already includes nearly 300,000 quasars whose spectral redshifts show two distinct maxima at z = 0.8 and the other at z = 2.15. If one interpreted this redshift as a distance, one would observe the state 875 million years after the Big Bang from Earth. So the light of the example quasar would travel for 12.8 billion years. The spectrum of Figure 7.7 shows a very strong, highly ionized carbon line, which is characteristic of quasars.

This shakens the notion that quasars are very early cosmic objects, because to get carbon from hydrogen, the entire Bethe-Weizsäcker cycle [7.30] should have passed through. A complete run of the cycle is said to be enormous periods of magnitude of hundreds of millions of years for massive stars, with the generation of oxygen from nitrogen alone estimated at over 320 million years. The accumulation of such massive masses (about 12 billion solar masses) over the remaining time of about 600 million years, as they should be accomplished solely by gravity, makes pondering even hard-nosed representatives of the standard model.

Even more puzzling than the extreme redshift of individual quasars is the observed flickering of their light. This flicker is not seen as a property of the source itself, but gases in the Milky Way should be responsible for [7.31].

7.4 The System-theoretical Model of the Cosmos

Already under section 3.5.1 How to overcome the Dilemma with Infinity? we saw a solution for a system without known initial conditions. Laplace has thought of the world as if there were no interaction among the individual objects. Einstein believed that time was independent of space and therefore of movement. We now want to consider the intergalactic space as an open dynamic system with the input *x(t)* and the output *y(t)*, *x* and *y* are respectively input and output for matter and energy, without making any agreements about these individual system elements. The system itself is a 'black box' and then we can define:

A state function $dz/dt = f(z,x(t))$
and a result function $y = y(z,x(t))$ with $z = z_1, z_2, z_3, \dots z_n$

z is the set of the current states in the moments of considering. Please note that each system has a lot of subsystems that are described in the same way. It does not matter if it's the cosmic network, a galaxy or a solar system, or a planet. This coupled pair of equations avoids two problems.

1. having to reckon with unknown initial values and
2. to consider time as an independent entity in this system.

However, there is another problem. How can I arrange the states in the subsystems? What time measure is on the intergalactic scale? The light year is not appropriate, as it is related to the earth. The cesium line does not occur either. Only the H_α-line of hydrogen can be seen in all galactic spectra and so in all subsystems. Due to the different redshift, however, all cosmic clocks tick differently. Let us remember how, in the Middle Ages, there were no mechanical watches that measured time. One burned off candles and measured the shortening of the candle. In galaxies, it is the consumption of hydrogen for the fusion of heavier elements, which are incubated from the hydrogen clouds. The loss of intensity of the hydrogen line at 656.3 nm thus becomes a measure of the age of a galaxy. On Earth, the height of the spectral line usually suffices. However, since the spectral lines in the galaxies have different widths, we use the area of the hydrogen line, the *equivalent width*[11]).

7.4.1 What the Spectra of the Galaxies tell us

Each light source has a characteristic optical spectrum in which the information about the light source is encrypted. Gases emit line spectra and liquid or solid bodies emit thermal continuous spectra. A light source like a star consists of a liquid magmatic nucleus and a luminous atmosphere. In this atmosphere, magmatic vapors diffuse and produce dark absorption lines in the spectrum of bright thermal light,as can be

11 The equivalent width of a spectral line is a measure of the area of the line in a ratio of intensity to wavelength. The equivalent width does not depend on the characteristics of the equipment used and is therefore well suited for comparing different measurements.

seen in Figure 7.8. They are named after Joseph Fraunhofer. In 1814 he discovered the spectroscope with which the light spectrum can be analyzed for wavelengths.

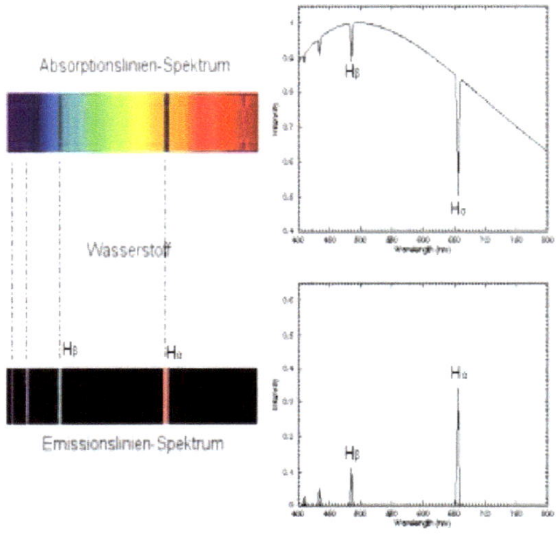

***Continuous spectra** arising from the temperature of a body. This spectrum is described by the Planckian radiation curve. Absorption lines due to colder gases are in the foreground.*

***Line spectra** arising as a result of excitation of electrons in the atomic shell in an atmosphere. The line spectra may result in absorption or emission depending on whether the background radiator is hotter or colder.*

Figure 7.8: Two types of line spectra from a spectrograph and a spectrometer

So far we have only dealt with the spectral redshift of the spectrum as a whole. Now we see the line intensity in the form of the area of a peak, which can provide information about the quantity of the substance in the plasma. The relationship between thermal background and line radiation gives us clues also. The Planckian radiation curve of the background radiation gives us information about the temperature of the radiator.

When we look at galaxies, their spectra do not have a typical Planck shape, because galaxies consist of many stars in different stages of development. Hubble has studied galaxies and classified them according to their outer shape. Look at figure 7.9.

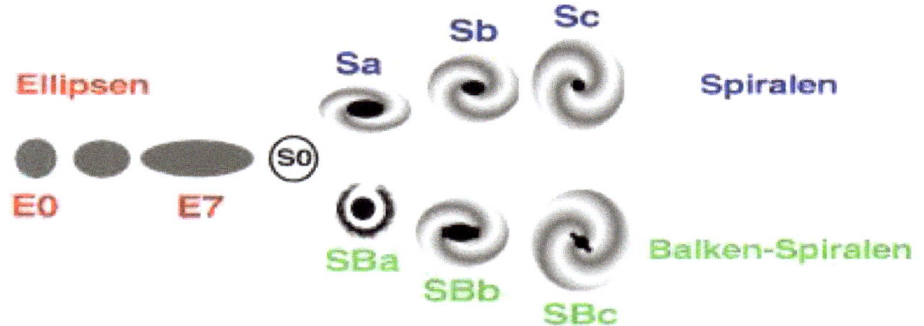

Figure 7.9: Hubble systematic - Reference:
https://www.spektrum.de/lexikon/astronomie/hubble-klassifikation/188

As mentioned in section 6.3, the Galaxy Zoo 1 project started in 2007 [7.32]. It consisted of a data-set of 900,000 galaxies taken by the Sloan Digital Sky Survey [7.33]. Of these, 667,945 galaxies were presented to the general public for classification into spiral galaxies and elliptical galaxies, and anyone interested could participate in the project. 28% of these galaxies were recognized as spiral galaxies and 9% as elliptical galaxies. 62% of the galaxies could not be safely assigned to one or the other class, because the judgment of the form is a subjective decision. Since I had participated in this project, I was also all the spectra available. When looking at the spectra, I came upon a system that could overcome the uncertainty in the classification. We will discuss this system in the following section.

7.4.2 The Evolutionary States of the Galaxies

Galaxies are not randomly distributed in space, floating as subsystems in huge Birkeland streams, separated from huge voids of extremely low consistency. Their formation was discovered by Antony Peratt when he simulated the union of two Birkeland streams in the computer. [7.34] Thus their formation is caused by external energy input. Thus, galaxies have a certain similarity with tornadoes in our weather system, caused by the warming of seawater.

But let us come back to the fusion of hydrogen and the classification of galaxies. As long as enough hydrogen is still available, the fusion chain will be maintained from the beginning and thus the emission lines of hydrogen will remain visible as long as hydrogen is replenished from the environment of the galaxy. The analogy to a tornado is that this as long as it moves over warm sea water, it catches more and more vapor supply. So galaxies collect hydrogen. At the same time, heavier elements are also produced in the galaxies, as can be seen from the spectra. The ionized gases nitrogen and oxygen are especially noticeable. So while the gases in the outer reaches of the galaxy produce emission lines, a growing galaxy nucleus provides for ever more intense thermal radiation, resulting from the formed stars, where the fusion continues to progress. According to the fusion rate, the radiation intensity of the spectral lines of hydrogen decreases with time, and the radiation intensity of the nucleus increases and, when the hydrogen is consumed, should have pass its maximum.

Both processes are therefore in opposite directions. From this both processes, conclusions can be drawn about the state of development and thus on the age of a galaxy, even if at present there is no reference measure for a time scale in years or rotations of the galaxies. We see only different states distributed over the cosmos, that we can rank. From quantitative spectroscopy we know that the intensity of a spectral line can give information about the amount of excited substance, if a calibration can be used to establish a functional relationship between the intensity of the spectral line and the amount of substance. If one were to know the rate of fusion of hydrogen, one could also calibrate a clock formed from the intensity of the hydrogen line. The Hubble systematic in Figure 7.9 shows three different types of galaxies. I suspected that these types could correspond to different stages of development.

To get reasonably accurate information from millions upon millions of galaxies, you need samples of the order of hundreds of thousands of galaxies. The sample considered here includes over 340,000 galaxies from a ring of 0° to 15° dec above the ecliptic. The database Release 7 [7.35] contains 924,000 galaxies. The sample comprises about one third of the total database, which was automatically screened. The most prominent emission line of hydrogen is the H_α-line from the Balmer series at 656.3 nm. This is the line that arises when an excited hydrogen atom falls back from the 3rd to the 2nd excited state.

This line can be found in every optical spectrum of all galaxies and it is central enough that it can still be observed well up to a redshift of z = 0.4. The limit of z = 0.4 is also due to the fact that, up to this limit, the determination of the distance from the decrease in the brightness of the Cepheids still correlates well with the redshift. Therefore, the intensity of the H_α-line should serve as a measure of the amount of available fuel in the galaxy and thus provide a relative comparison of the age evolution of galaxies.

But this one line is not enough. Another conspicuous hydrogen line is the H_β-line at 486.3 nm, which is emitted when returning from the 4th excitation state to the 2nd excited state. It can be either stronger or weaker than the H_α-line. Since more energy is needed for their stimulation, this line will not be well observable across all ages. However, we can compare the thermal continuum at this point with the continuum at the location of the H_α-line.

This method has the advantage over the color analysis with the green and red filters that this criterion moves with the red shift and thus produces no distortion of the conditions towards blue with changing redshift. The H_α-lines and the thermal continuum at these two points formally distinguish 3 basic spectral types into which galaxies can be classified.

Since there are no benchmarks for exact age information and it is also unknown how the emission behavior of the gases in the galaxies influence each other and thus can falsify the result, we will work with three main age classes or states. Each of the main classes is divided into decadal subclasses according to the H_α-line intensity. In the first class, the maximum of the thermal radiation is in the UV range. That corresponds to a temperature of about 12.000 K. The second class and third class form their maximum radiation in the visible spectral range, which corresponds to an average temperature of 6,000 K. In the third class, however, the emission lines are transformed into absorption lines. This fact is reflected in the following table.

Spectral type	Criteria
I : juvenile galaxies	Continuum at H_β > Continuum at H_α and H_α > 0
II: mature galaxies	Continuum at H_β < Continuum at H_α and H_α > 0
III:ancient galaxies	Continuum at H_β < Continuum at H_α and H_α < 0

Table 7.1 Rules for spectra classification

For our measurements, we do not use the height of the spectral line, but the equivalent width, which is already stored in the database for each spectral line. From the difference between two sufficiently distant wavelengths of the thermal background radiation of the galaxy nucleus, it can be seen whether the intensity maximum is shifted towards shorter wavelengths or towards longer wavelengths. For this we use the continuum at the H_β-line. Thus we can still compare the hydrogen supply of the galaxies even with larger red shifts of the spectral lines. For the evaluation of the spectra only areas could be used, which remain visible despite redshift over all spectra. This is possible up to a redshift of 0.4, which corresponds to a distance of about 1,714 Gpc, or 5.55 billion light years.

Spectral type	Number of galaxies	Total share based on Sum	Share up to z = 0.15 in relation to the number of Spectral type
I	41 971	12.29%	89.46%
II	230 667	67.5%	91.65%
III	68 857	20.16%	61.64%
Sum	341 495		

Table 7.2: Overview of the result of sample classification

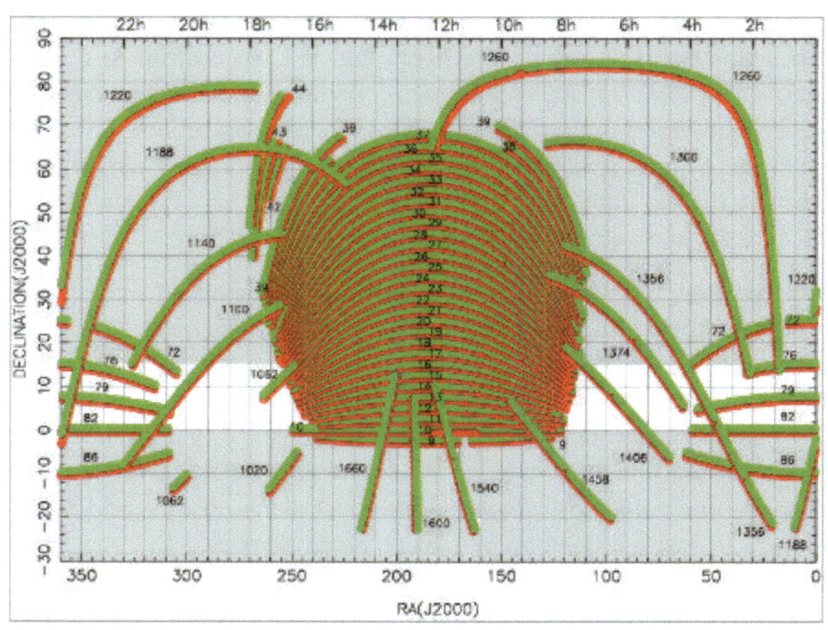

Figure 7.10: Sample in relation to parts of the sky covered by SDSS
Reference: SDSS Release 7

A section of the sky has been scanned that corresponds to the white area in Figure 7.10. The green-red bows show the areas of the sky recorded by SDSS. In this window, exactly 341,495 data sets of galaxies were found that met the above conditions. They are distributed as follows over the 3 spectral types: Table 7.2 says that although the proportion of young galaxies is only about 12% of all galaxies, 89% of them are found at a distance of less than 2.1×10^9 light-years. However, the distribution on the spectral types is not uniform. From type III are only 62% of the galaxies in the given range.

How can you interpret this result? First you have to know the limits of the telescope. A 2.5 m reflector telescope has a theoretical resolution of about 0.046". However, the reception of the signal is limited by the photo-sensitivity of the detector. The question is: how bright must be a galaxy that it can still be perceived from a certain distance? SDSS can detect objects up to a relative brightness of 23. With increasing distance, the fainter galaxies are separated out, so that up to z = 0.15 (equivalent to 645 million light-years) only more than half of the galaxies can be found. In the first two classes it is even around 90% of the galaxies. In fact, it is the Sloan Great Wall, the largest structure that was discovered until 2007. It lies at a distance of z = 0.1 between 130° $< RA < 230°$ and falls exactly in the sample selected here.

The Big Bang thesis is not supported by this finding. In an explosion, one would observe an approximate uniform distribution of the dissipating matter. Since the redshift in the Big Bang model is interpreted as a Doppler effect, it can be observed in all directions, so we would be at the center of this explosion. Since this is unlikely, the volume of the cosmos would have to expand. It is also claimed that with increasing distance one could see into the further past of the cosmos and observe more young galaxies. None of this can be confirmed.

Let's take a look at the distribution of each type of galaxy in more detail. It should be remembered that the measure of aging is the reduction in the proportion of hydrogen in the galaxy, because as a result of the nuclear fusion processes, this proportion must decrease in favor of heavier elements. Practically it will be merged to bigger atoms with time. The galaxies collect in their center more and more mass in the form of heavier atoms and thus increases in the interior of a galaxy their density, which emerges as a galaxy nucleus in our photos.

The onset of fusion causes an electron deficiency, as we have already seen in Chapter 5 from the droplet model. The lines of ionized oxygen and nitrogen bear witness to this. Under the increasing influence of electromagnetic forces, this nucleus of young stars will continue to grow until it over-shines the again declined electrically excited plasma. The electrical phenomena in the galaxies are nothing other than what we observe on Earth in the polar regions as polar light or aurora borealis

This means that the mean developmental period lasts at least 5 times longer than the youth time of the galaxy. But that is only possible if hydrogen is replenished constantly. This can be explained as follows. Due to its growing power, the grown nucleus is capable of sucking unseen molecular hydrogen like a vacuum cleaner into its rotating arms from all over the neighboring environment, splitting it into atomic hydrogen and ionizing it. Molecular hydrogen would also be a reasonable explanation for the so-called dark matter, as Paul Marmet [7.36] stated, because it is extremely transparent in the optical domain and only detectable in the infrared.

Type I. Young Galaxies: In the galaxy spectra, In addition to the Balmer series of hydrogen, the strong lines of triply ionized oxygen at 500.8 nm and doubly ionized nitrogen at 655 nm and 658.5 nm are also striking, suggesting strong electromagnetic fields in the galaxies, as known from satellite researches in the near-Earth space. Since the electromagnetic forces are 39 orders of magnitude stronger than the gravitational forces, the gravitational forces in young galaxies hardly play a role, which would also explain the almost constant rotational velocities over the variable radius of the galaxies, as Peratt [7.37] calculated in 1987. This indicates that galaxies are cosmic eddy currents forming from cosmic plasma threads.

The 41,971 spectra of the young galaxies are distributed as shown in Figure 7.11, pointing to the given 4 subclasses, corresponding to the equivalency width of the H_α-line of hydrogen (*ewHa*). We observe here a skew-symmetric distribution curve with a maximum of mass at z = 0.1. This maximum is flanked by particularly active galaxies. 90% of all young galaxies found, are at distances less than z = 0.15. Then the brightness of the galaxies decreases significantly, as can be seen on

the light blue columns. However, the turquoise pillars, which is low in hydrogen gas, show from z = 0.3 an increase against the trend. This may be related to the fact that the galaxies are significantly weakened

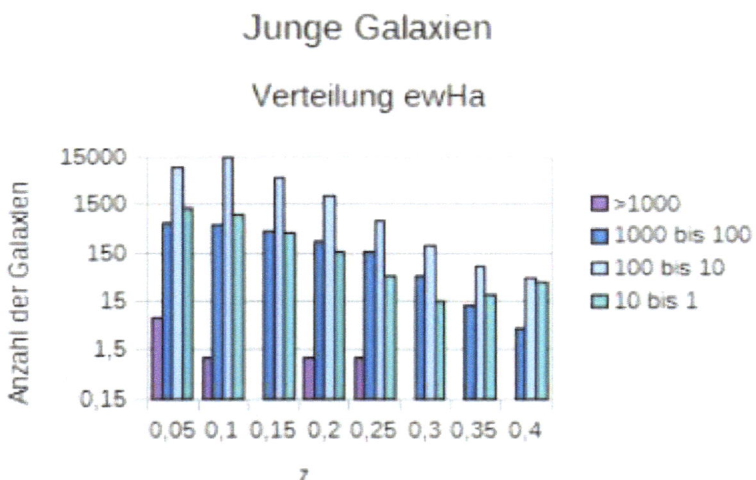

Figure 7.11: Distribution of young galaxies correspond to their hydrogen supply

in their intensity and thereby slip into the lower 'intensity class'.

The light of the galaxies then becomes increasingly weaker, so that under constant recording conditions only very bright galaxies are detected by the telescope. In addition to the Balmer series of hydrogen, is interstellary seen triply ionized oxygen and triply ionized nitrogen. The importance of these gases is not yet clear, but where there is much oxygen, the nitrogen content is lower and reversed.

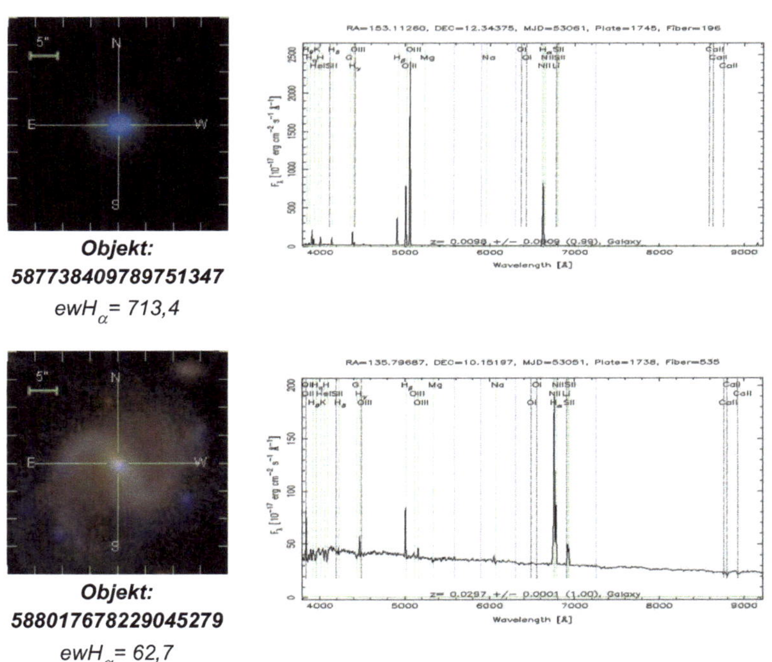

Table 7.3: Galaxies of Type I - Object 588017678229045279 shows the typical beam structure.
Reference: http://skyserver.sdss.org/dr14/en/tools/explore

The assumption that we have to find more young galaxies at a greater distance is not correct for the present sample.

Type II. Mature galaxies: The largest number of galaxies are those with pronounced thermally glowing nuclei whose optical background radiation forms a pronounced background plateau around the H_α -lines, without reaching the H_α -line in height, and more or less to theH_β -line falls steeply.

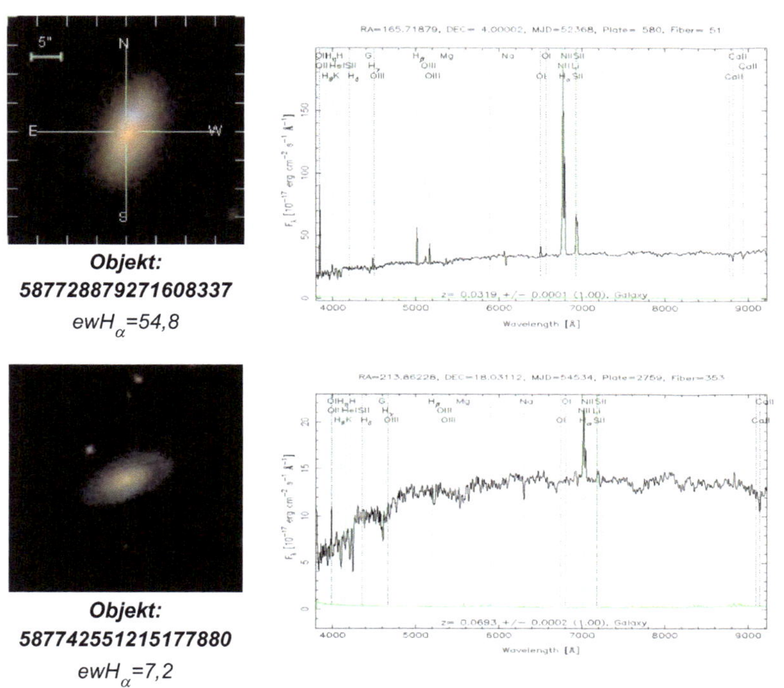

Table 7.4: Galaxies of Spectral Type II
Reference: http://skyserver.sdss.org/dr14/en/tools/explore

The pronounced lines of ionized oxygen and ionized nitrogen have decreased compared to type I, in particular the triple ionized oxygen has been reduced in favor of the double ionized oxygen, which indicates a reduction in the electromagnetic forces. This means that fewer and fewer gases can be collected from the distant rooms of the universe.

However, elements of higher atomic numbers such as sulfur are now being created. The distribution of the H_α-lines of the galaxies over the 5 intensity classes is shown in Figure 7.12. The intensity has decreased somewhat compared to the young galaxies, but the proportion of those further away is somewhat larger.

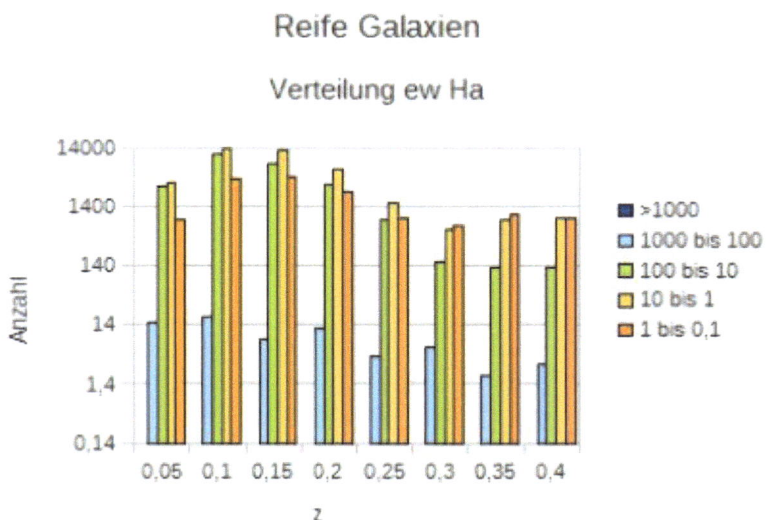

Figure 7.12: Distribution of mature galaxies according to their hydrogen supply

Type III. Old galaxies: The old galaxies are characterized by having a pronounced thermal background spectrum with hydrogen spectral lines in absorption and other absorption lines such as sodium, ionized magnesium and calcium. The electric fields take a back seat, as the fusion consumes more energy than it delivers in the range of larger atomic numbers, as can be seen from the extensive absence of lines of ionized elements. Due to the absence of strong electromagnetic forces, the spiral structure dissolves, revealing the typical character of the elliptical galaxies. By mass concentration in the center of the galaxy, the gravitational forces may have increased further against the electromagnetic forces.

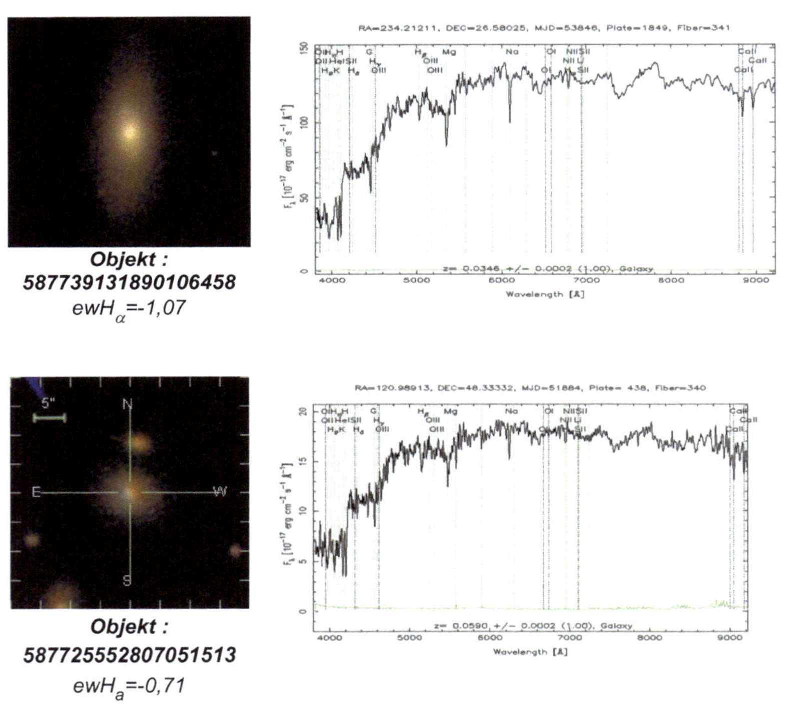

Table 7.5: Ancient galaxies of spectral type III
Reference:http://skyserver.sdss.org/dr14/en/tools/explore

Figure 7.13 shows the distribution of old galaxies, with the older ones now being characterized by the front brighter intensity bars. Looking at the class distribution, it is noticeable that among the 68,857 old galaxies, 53% of the spectra are in the intensity range from -0.01 to -0.1 and 46% in the intensity range from -0.1 to -1.

Only less than one percent of the spectra shows deeper H_α-absorption lines, indicating that the remainder of hydrogen is rapidly consumed at this stage. For the old galaxies, the maximum of all galaxies is in the second and third range classes. The minimum at z between 0.25 and 0.30 can also be seen here, though not so pronounced. Also amazing is the relatively flat distribution curve in the intensity class -0.1 to -1 and the counter-running distribution in the intensity class from -1 to -10.

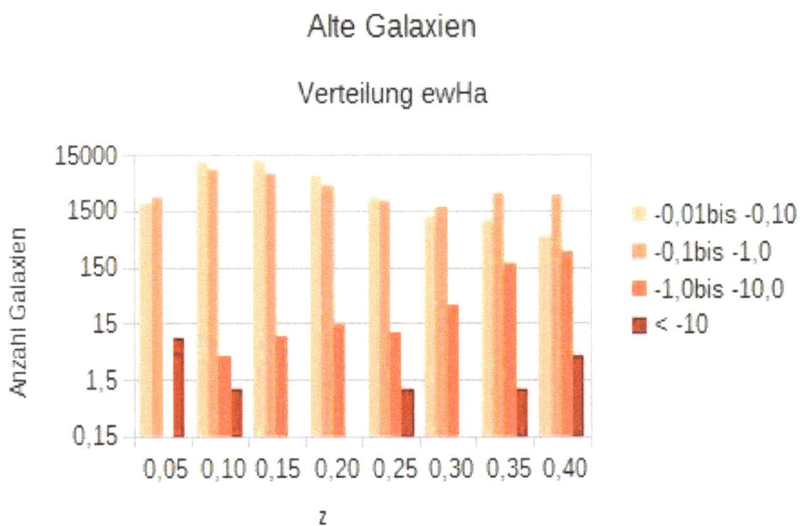

Figure 7.13: Distribution of ancient galaxies according to their hydrogen supply

In summary, young and mature galaxies are predominantly found in the Sloan Great Wall, whereas the ancient galaxies are more evenly distributed throughout the space, and there has to assume a light attenuation to the blue end of the spectrum depending on the distance in the cosmos. From the frequency of distribution, it can be concluded with steady hydrogen burning, that the youth stage as well as the old stage of development each make up about between 10 and 20% of the total number of the control sample. The mature galaxy stage, however, exists over 70% of the entire lifetime. How fast an old galaxy burns out when the hydrogen supply is used up can not be answered. The distribution minimum at z between 0.25 and 0.3 indicates a large-scale inhomogeneity.

7.4.3 The Velocity distribution of Stars in Galaxies

The rotation speed of galaxies has been studied and compared with the law of gravity. As shown in Figure 7.15, the rate of rotation towards the edge barely decrease, although according to the law of gravitation in our planetary system, it should have followed the blue Kepler curve. Standard astronomy therefore assumes that there must be a ringlike cloud of a large amount of dark matter around the galaxy to explain the rotation speed. This is accepted today in astrophysics as part of the standard model. But nobody knows anything about this mysterious dark matter, but it should interact with the galaxy by its gravitational force. Is there perhaps a better explanation for the velocity distribution of the masses in a galaxy, without having to resort to mysterious phenomena?

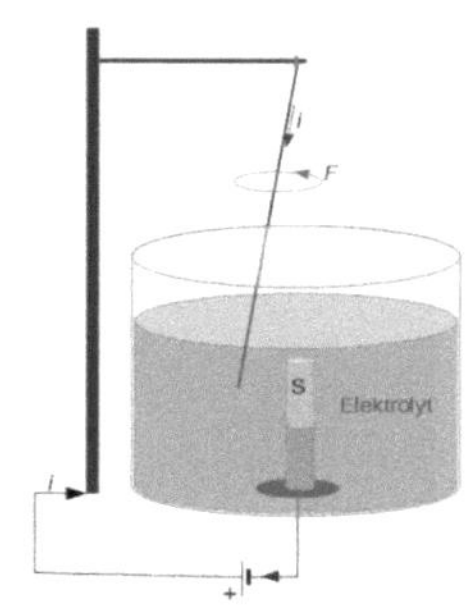

Figure 7.14: Principle of a Faraday engine

Michael Faraday discovered around 1821 that a metal wire that plunges into an electrolyte in which a magnet is located begins to circle the magnet when a voltage is applied to the wire and electrolyte (Figure 7.14). Now you can think of the magnet as being replaced by a ring current. Flow through a plasma cloud two Birkeland currents that interact with each other, then the plasma cloud acts like an electrolyte and it also has magnetic properties. The result is that a vortex forms. This is exactly what Antony Peratt simulated on a super computer and found that the angular velocity to the edge of the vortex remained fairly constant, since, according to the law of induction, the angular velocity, depends on the quotient of the voltage to the product of the magnetic flux times the transversing surface, and not on the radius of the galaxy.

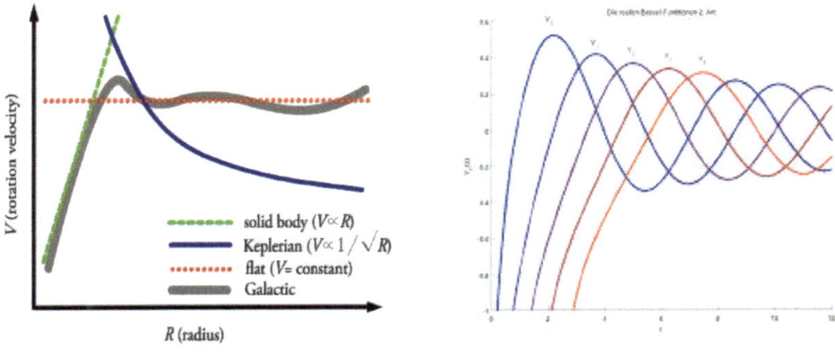

Reference: WIKIPEDIA   *Reference: mathe.tu-freiberg.de*

Figure 7.15: Rotation speed of the stars in the Milky Way compared with the Bessel function 2nd type

If you compare the speed of rotation of the stars in Fig 7.15 with a Bessel function 2nd type, you can see the similarity in the radial speed distribution. Because of the low resolution in the galaxy, the ripple is less than would be expected in theory. This clearly identifies the galaxy as an electrically powered entity and contradicts the assumption that a halo of dark matter would exist around the galaxy because the velocity does not follow the blue Kepler line. Since the middle of the last decade of the 20th century, measurements on the H_{α}-line have resulted in more and more galaxies not only with a wavy plateau (gray line), but even with areas of counter rotation[12]) Don Scott explains the mechanism in [7.39].

12 E. M. Corsini - Counter-Rotation in Disk Galaxies https://arxiv.org/abs/1403.1263

7.5 The Spectral classes of Stars in our Milky Way

7.5.1 The Star spectra

Star spectra are quite different from galaxy spectra. They are continuous spectra, which are describing with the Planckian radiation curve, reaching a maximum characteristic for the temperature of their surface. This maximum determines the color of the star.

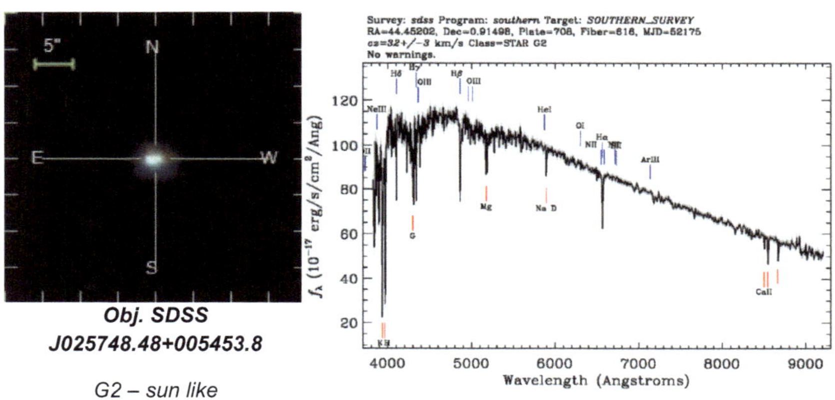

Obj. SDSS J025748.48+005453.8

G2 – sun like

Table 7.6 Sun-like star
Reference: http://skyserver.sdss.org/dr14/en/tools/explore

At first glance it is striking that there is hydrogen in the sun-like star atmosphere, but helium is missing in the yellow stars.

There is a lot of calcium, magnesium and sodium for this. Helium can only be found in the blue stars of classes O and B. This contradicts the classic solar model as a thermonuclear furnace that burns hydrogen to helium in order to explain our solar energy.

7.5.2 The Classification of the Stars

The most reliable information about the stars is obtained from the analysis of their radiation. Since Fraunhofer 1814, the absorption lines in the solar spectrum and in the spectra of the other stars are known. In addition to H and He, the elements Na, Mg, Ca, Ti and Fe were also found. Stars are classified in the Hertzsprung-Russell diagram in the classes O, B, A, F, G, K, M according to Harvard, like in Figure 7.16 shown. More than 90% of the stars studied, are located on the main axis of this diagram, including our sun. Other than that, there are three isolated areas that can not be classified in it.

If you look at the star spectra from the SDSS database [7.40], you will find even more elements. Surprisingly, not all elements are found in the

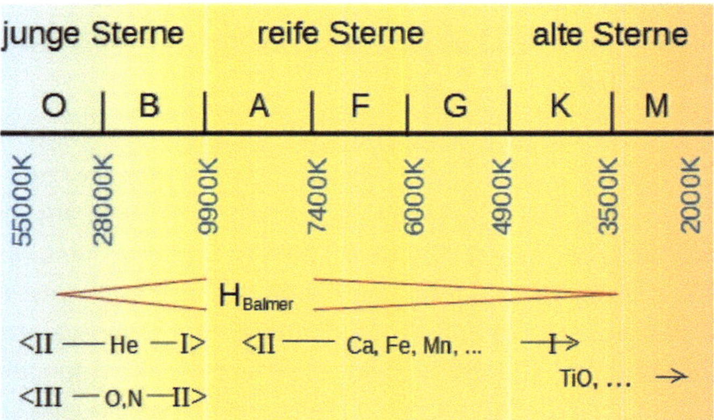

Figure7.16: Spectral classes of stars according to Harvard

stellar atmospheres. This applies to both light and very heavy elements. While the Balmer series of hydrogen in the spectral classes can still be seen in absorption up to star class F, helium can only be seen up to class B. Hydrogen, however, can be found a lot between the stars. However, obviously it is not distributed evenly.

As Figure 7.16 shows, no He-I lines are found at surface temperatures of stars below 10,000 K in star atmospheres at 402.6 nm or 447.1 nm. To ionize helium, 25 eV or 290,100 K in the star atmosphere is necessary. To find the helium lines He II at 468.7 nm and 541.3 nm, temperatures over 30.000 K at the star surface are necessary respectively 638.000 K in the star atmosphere, which is why the helium inside the stars was assumed as the result of a thermonuclear transformation. But this idea does not fit into the concept of an open system that receives its energy from outside.

7.6 The Solar Electric System

How wide is the solar system?

- Does it end behind the last planetary orbits?
- Does it end where the gravitational influence ends with the hypothetical Oort cloud?
- Does it end at the heliopause, where the solar wind is no longer perceptible?

This question can not be answered with sincerity. We know that the next star is four light-years away. Maybe then the radius of our system is about one and a half light-years. According to WIKIPEDIA, the central star is said to be responsible for 99.86% of the total mass of the solar system. The rest of 0.14% is then distributed to the planets, comets, dust, gas and cosmic plasma. If we compare the solar system with a gas discharge lamp, the mass ratio is similar. Also in spectroscopy, stars are identical to the surrounding plasma in our electrically operated gas discharge lamps. As early as 1893, Nicola Tesla demonstrated the importance of wireless electrical energy transmission and made gas discharge lamps, blaze without any wires.

If we look at the scheme of a gas discharge lamp, we can identify the plasma as yellow and blue light in Figure 7.17. The Birkeland current is available due to its low density as dark current over the longest distances. Only when the density increases, the gas does begin to glow. It forms a positive pillar. Immediately at the anode, the light becomes particularly intense. Remember the electric atom model that we discussed in section 5.4. The Droplet Model of the Atomic Nucleus. There we found that nuclear fusion has an increased electron consumption to stabilize the nucleus. Consequently, stars are anodes. The distant cosmos is then the cathode.

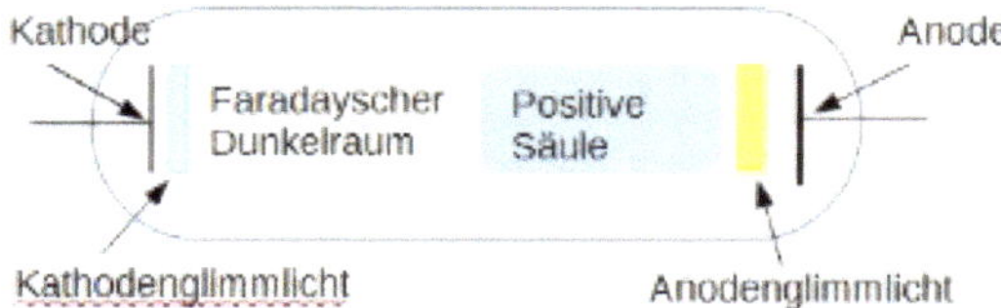

Figure7.17: Low-pressure gas discharge tube. Reference: WIKIPEDIA

A typical phenomenon of a positive column is the tail of comets, which forms especially in perihelion, i.e. where the region of the positive column is. Comets come from the depths of the solar system on an elongated elliptical orbit from an area with a negative potential to an area with positive potential. The idea of electric comets has a long history, as Hannes Täger [7.41] proved. It is as long as the story of the Enlightenment and dates back to Hugh Hamilton in 1769. He described both as cosmic electrical phenomena in an essay on the Aurora borealis and the comet's tail.

Another argument for an electric solar system is the distribution of the torques of the planets in the solar system. E. J. Lerner cites the problem with the moments in his book *The Big Bang Never Happened* [7.42] to show how the plasma model can explain the distribution of torques in the solar system, which was not possible in the standard model.

> *»The sun has a very small rotational speed compared to the planets due to a very small torque. Since Laplace, it has been suggested that the planetary system is condensed from a single interstellar nebula. But there was a difficulty: for every isolated object the total momentum - the product of radius, velocity and mass - is constant. This means that as the radius narrows, as the rotational speed increases. (The pirouette during ice skating demon-*

strates this impressively.) So the fog had to rotate faster during the condensation.
Given the known mass and orbit of each planet and the sun and their rotational speeds, it is easy to calculate the momentum of the solar system and determine which planets have the greatest torque around the sun. The result is confusing. If the sun had retained the torque of the overall system, it would spin itself once in 14 hours. But one turn of the sun takes 50 times longer, about 28 days. The sun has only 2% of the total momentum of the solar system, while Jupiter with one-thousandth of the mass has 70% of the total momentum. Thus, a transfer of the moment of the sun to the planets must have taken place. On the other hand, since the neighbor sun is about 4 light years away, the cloud from which the solar system is condensed must have been at least one light year in radius. If the cloud had completed a revolution at the same speed as the galaxy, then a torque would come out that would be 700 times the torque of today's system. Such a cloud would not have formed a star if it had retained its torque. It would rotate faster and faster with further contraction until the contraction by the centrifugal forces would have come to a halt at about $20x10^9$ km. However, if the contraction process had left only 2% of that moment, the process would have come to a halt at about 10^6 km, more than a dozen times the size of the sun and far too big for a star. The gas would be far too cold to burn hydrogen to helium. In short, to form the solar system, it had to lose 99.9% of its initial torque and transmit 98% of the rest of the moment to the planets. How could that happen?
H. Alfvén was convinced that this could have caused the electrical currents generated by the protostar's magnetic field. Suppose a rotating magnetic body is surrounded by clouds that do not rotate so fast, then the magnetic field of the central body will rotate with it and pull through the clouds, inducing an electric current. This current will entrain the cloud in the direction of rotation of the magnetic field. Like a big blower, the magnetic field transmits the torque to the clouds. As the gigantic stream accelerates the cloud, it will slow down its momentum as it returns to the sun.« End quote!

The heliospheric positive current has been detected since by many cosmic probes, such as Ulysses and Voyager-1 and 2, and is now reported and published daily in density and speed by the Solar and Heliospheric Observatory (SOHO) via https://sohowww.nascom.nasa.gov/ and http://spaceweather.com. According to the solar activity it is subject to statistical fluctuations. With our knowledge of the electric droplet model of the atom, we suspect the solar current as the ejection of surplus protons in the nuclear fusion at the solar surface.

7.7 The Double layer

Electrical voltage is a consequence of charge separation and any charge separation is accompanied by an electromagnetic force field maintained by chemical, thermal or mechanical forces. The double layer is an example of an open system far from thermal equilibrium. We know many types of charge separation. Phase boundaries always play a role. These prevent the charge equalization. So you can build batteries and capacitors. We have found many technical applications of charge separation in the last two centuries. On Earth, we have natural charge separation in the ionosphere and storm clouds. Why we do not want to admit the universe that possibility?

The luminous structures are nevertheless a clear testimony to phase boundaries in which there is a balance of charges that may have been separated elsewhere. We have to admit that we have not fully understood the physical processes, because standard physics speaks of

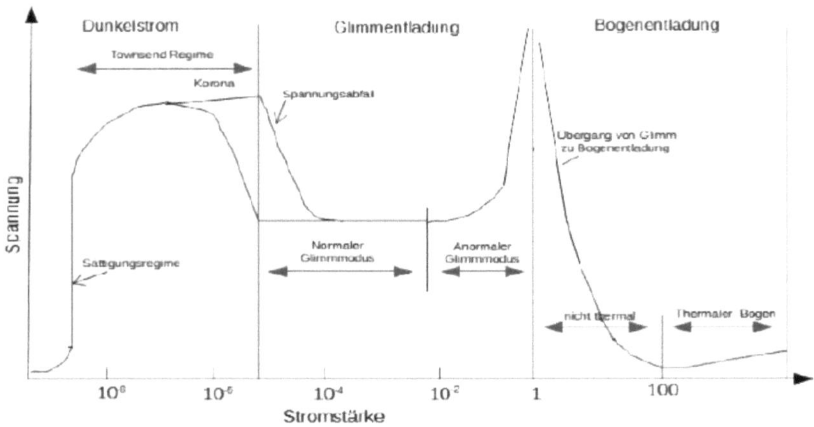

Figure 7.18: Electric discharge regime Reference: WIKIPEDIA

quasi-neutrality in nature. This quasi-neutrality does not explain the transmission of electricity over long distances, neither the technical nor the transmission in cosmic plasma flows.

So far we have considered the plasma current only qualitatively, so let us consider it semi-quantitatively now. It is called plasma, when it is a gas with at least 10% ions, which is traversed by faster electrons. If electrons in a thin plasma flow much faster than the ions, the ions cannot capture the electrons. We then speak of a dark current. If the speed difference between electrons and ions falls below the detachment speed, there is a possibility of electron capture. The transition to glow mode is associated with a voltage drop. The glow of the plasma comes about through recombination with the trapped electrons. Before an arc discharge, the voltage rises again sharply. A so-called *double layer* is formed by charge separation on neutral particles. Only when the breakdown voltage is reached, ignites a flash or arc that discharges the double layer. If one observes the voltage dependence of the current, one finds the following characteristic, as shown in Figure 7.19.

A Double-layer was first described by Hermann von Helmholtz in 1853. From him comes the term. He wrote:

> *»In the following, I shall always understand by an electric double-layer only those two layers which lie on the opposite sides of a surface at an infinite distance, and one of which contains as much positive electricity as the other negative.«*[7.41].

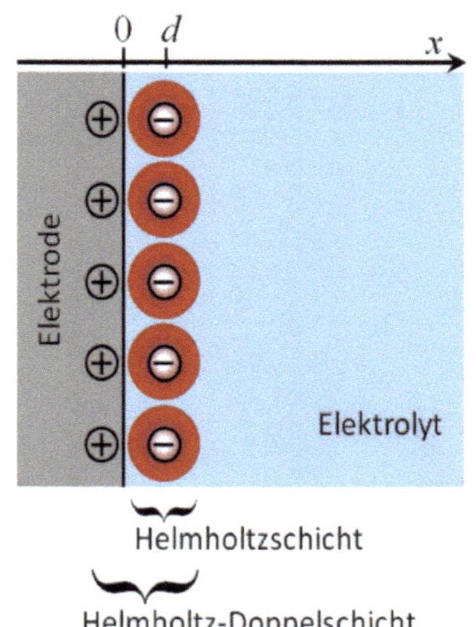

Figure 7.19: Double layers by Helmholtz
Reference: WIKIPEDIA

We observe electric discharges in thunderstorms as lightning, as auroras, as comet tails and as solar flares. But precisely these double layers are the result of the movement of matter and the cause of the electrical discharges. We are just beginning to understand the meaning of the double layers for cosmic evolution. The consideration of the cosmos as a quasi-neutral system is not very helpful because the subsystems are not neutral at the moment. These are coupled open systems with mass and energy exchange.

7.8 Criticism of the Classic Sun Model

The Sun's standard model says the Sun is a thermonuclear fusion reactor that fuses hydrogen inside to helium. The main constituents of the sun would be hydrogen 73%, helium 25% and the remainder C, N, O, ... 2.4% would come from meteorites. As recently as 2005, research information from the Max Plank Institute reads:

> *»The core of the sun is the source of solar energy and the powerhouse for the entire solar system, including life on earth. At temperatures of 15 million degrees and a pressure that compresses the predominantly hydrogen gas to 13 times the density of lead, conditions are just right for nuclear reactions to occur. At these temperatures, most of the atoms are broken down into their constituent protons, neutrons and electrons, and because of their high density and energy often clash at high velocity.«* [7.44]

With some knowledge of kinetic gas theory, we have to ask the question: How should 15 million degrees of temperature come about if the atoms have no freedom of movement? And further on we read there:

> *»The proton-proton reaction chain is the dominant reaction in the solar interior. It leads via the intermediates deuterium (D) and beryllium (Be) and with emission of gamma quanta and neutrinos to the construction of the noble gas helium (He), which accumulates as 'slag' of the nuclear reaction in the center of the sun.«*

Why should helium be stored inside the sun? Has anyone really measured that, that he can claim it? No its pure faith. And the existence of neutrinos responsible for transporting heat to the surface is in conflict with the NIST data. Another argument against the theory of the thermonuclear furnace is the fact that the electric forces are much stronger than the gravitational forces. If the sun were such an oven, it would explode like a hydrogen bomb under these circumstances. Consequently, the nuclear fuel must be supplied from the outside in small doses. In the spectra of the active galaxies we have found enough hydrogen between the stars in section 7.4. There is absolutely no information about the interior of the sun, even if seismic measurements are claimed. However, if calcium and iron are detected on the surface in the solar atmosphere, then it is obvious that there must be a magma melt inside the sun, as on Earth, that develops a certain vapor pressure. The assumed mean density of the sun is 1.408 g/cm³. The density of calcium is 1.378 g/cm³ at 1,115 K. Boiling point is at 1,757 K [7.43] at a pressure of 1 Bar. Pressure and temperature increase counteract the density. The much higher pressure on the sun could keep the calcium fluid at 6,000K. Therefore, the density of the sun does not contradict the idea that it consists essentially of molten calcium, while the iron content must be rather low because of the low density.

Usually we can not observe the surface of the sun because it is obscured by the photosphere, a shell of hot plasma. »Here we could actually expect the sun to 'end' if the distribution of internally generated energy were not fundamental to maintaining its mechanical equilibrium, as claimed by the accepted theory.« wrote Ralph E. Juergens in 1979 [7.46]. But explaining the granularity of the photosphere as a result of

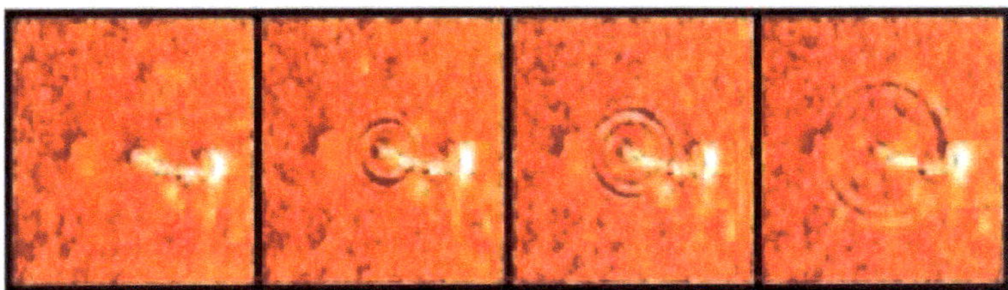

Figure 7.20: Waves on the sun surface Reference:SOHO Mission NASA

heat convection from inside the sun failed miserably. He identified the granules as anode tufts, as observed by Irving Langmuir in 1924 during plasma experiments in the laboratory [7.47]. These tufts consist of positively charged atoms that fuse into threads. In the meantime, the tufts have been recognized as a dissipative structure[13]) in an open system [7.48].

Only when the photosphere tears open, what happens through the sunspots, you can take a look at the surface. But then you can get such images, as shown in Figure 7.20 as a series. You can see significant concentric surface waves, indicating a glowing melt.

If the sun's surface were of hydrogen, it would not have a thermal spectrum, but a line spectrum of emission lines, and you would see a faint light of red and blue filaments, as in the plasma lamp of Nicola Tesla, or glowing plasma clouds like precursors of galaxies. The hydrogen is not in the stars, but in between, as the spectra of galaxies show. In that sense, one can not look at the sun in isolation from its surroundings.

7.8.1 Why is the Surface of the Sun Colder than the Solar Atmosphere?

The atmosphere of the sun is divided into three distinct areas, the photosphere, which extends to a height of about 600 km above the sun's surface and where the temperature drops from about 10,000 K to 5,000 K. The spectrum of the photosphere provides a thermal spectrum in which a variety of absorption lines can be found, arising from chemical elements that must be located in the overlying chromosphere, as they appear as absorption lines against a light thermal background. In the chromosphere, a lot of discharges occur, called spicules, which raise at a rate of about 150 km/s. In the chromosphere the temperature rises again to about 10,0000 K.

13 The term comes from Ilya Prigogine. These structures only form in open non-equilibrium systems that exchange energy, mass, or both with their surroundings.

Above the chromosphere, the corona joins. The transition to the corona at a height of about 2,000 km above the solar surface is associated with a temperature jump to about 2.5 million degrees. At these temperatures, the lighter elements till beryllium lose all of their shell electrons and they no longer appear in the optical spectrum because they are fully ionized. High temperatures mean high kinetic energy from the swirling protons, deuterons and other nuclei, which favors their fusion in collisions. The protons, which have not found any electrons, are accelerated towards Earth and are detectable as solar wind. There are two types of solar wind, the fast and the slow one In May 1999, NASA observed that the solar wind suddenly stopped for two days [7.49]. What could be the cause that there should be no solar wind from within the Sun, if it may never come from within the sun, but is an outward appearance? When the sun comes up with a positively charged particle stream, it can only mean that the sun itself has a positive potential with a corresponding field in which the protons are accelerated. You remember, Electrons are consumed in the nuclear fusion, which here are extracted from the interstellar hydrogen. The solar wind are the unfused hydrogen nuclei that has been stripped of its electrons in the corona. When the solar wind stops, it means that no hydrogen is currently entering the solar atmosphere anymore.

Since Stefan Boltzmann we know the relationship between kinetic energy and temperature. So you can quickly determine the temperature from the speed of the solar wind. At a mean velocity of the solar wind of 373 km/s, a proton temperature of 5.6 million degrees Kelvin is obtained.

Why are there sunspots on the solar surface and why does the corona change depending on the sunspot frequency? Note also that the corona is not uniformly bright, but concentrates in a loop around the solar equator. These bright loops are located within strong magnetic fields called active regions. Sunspots are located within these active regions. Solar flares occur in active regions. At the poles, the corona disappears in the X-ray image almost completely, as shown in Table 7.7.

All these effects can not be explained by the standard model of the sun. However, it explains the thermal radiation caused by the fusion of hydrogen into helium. But why should the hydrogen reservoir be within the Sun, where there is enough interstellar hydrogen, as we found in the spectra of the galaxies? You cannot look at the sun in isolation from its cosmic environment. It is integrated into the rotation of the galaxy, from which it draws its energy and which sweeps it away.

7.8.2 Classic Experiments for Modeling the Sun

One of the central questions in our time when replacing fossil fuels in the energy industry, is the question of how works our sun.

As long as we cannot answer this question, we will not be able to use their fusion energy. Can we recreate a small sun on earth? Based on the classic solar model, which assumes that the sun is a fiery gas ball, inside which hydrogen fuses to helium and the heat is transported to the surface by neutrinos, Andrei Sacharov started his first experiments at the Moscow Kurchatov Institute in 1952 for technical nuclear fusion. His principle was the inclusion of a plasma ring in a magnetic field, which should be heated up in the magnetic field enough to start the thermal nuclear fusion. The idea was that after the ignition, the fusion would continue by itself, as had been observed with nuclear fission. Therefore, the experiments were designed for pulse opera-

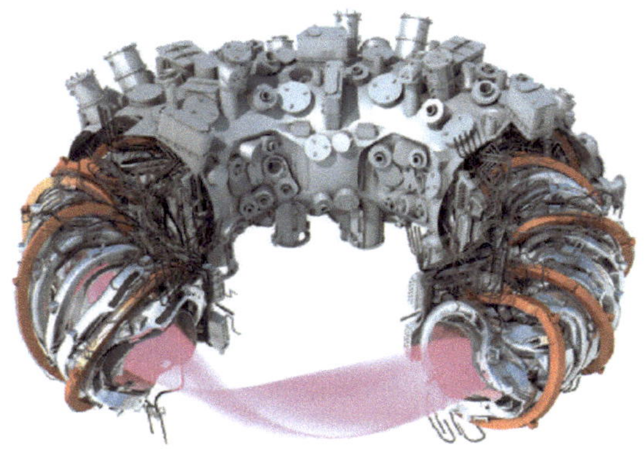

Figure 7.21: Fusion experiment Wendelstein 7X - Reference: Max Planck Institut für Plasmaphysik Germany

tion. This principle was given the name 'Tokamak'. It is short for „тороидальная камера в магнитных катушках“ (toroidalnaja kamera w magnitnych katuschkach) translates toroidal chamber into magnetic coils. However, there were two problems with this principle, the first was that the Lorentz force swirled the plasma and the second the chamber walls cooled the plasma. The swirling was finally countered with a complicated magnetic field design that was modeled on a multi-twisted Möbius tape. This created a helical magnetic field. This principle was called the Stellarator. However, more than 60 fruitless years had to pass before the new principle was decided.

Finally, on December 10, 2015, a short flash of light in the form of a helium plasma was generated for the first time on the Tokamak Stellarator Wendelstein 7-X fusion generator in Greifswald. To perform this experiment were required nine years of construction, in which several hundred tons of material were used. The project cost over a billion euros. The Max Planck Institute for Plasma Physics issued a press release on June 25, 2018:

> ***Stellarator record in fusion product***:*»Wendelstein 7-X has now set a record. Because it never reached previously measured maximum values for the so-called fusion product. This product of ion temperature, plasma density and energy confinement time indicates how close you get to the reactor values for a burning plasma. In the test runs in 2017, the plasma in the reactor was heated to around 40 million degrees ion temperature and had a density of* $0.8x10^{20}$ *particles per cubic meter.«*

Up to 75 megajoules of heating energy were used for the heating, and the plasma survived 2 seconds. However, there was still no sign of fusion. Wendelstein 7-X is currently out of service and is being renovated. All in all, a great technical achievement, as you would think, only our sun is a G2 star and there are just 5.600 degrees on the surface and there are no traces of helium in its atmosphere.

We asked a lot of questions about the standard model of the sun that it can't answer, which is why we discard it and move on to a different design.

7.9 The Electrical Anode Model of the Sun

We have asked a lot of questions to the standard model of the sun that it can not answer, so we reject it and turn to another design. Looking at the Milky Way, we see that the stars are embedded in a milky glowing gas, the main component of which is hydrogen. This glow is weak. It's just a glow. When masses glow, this aggregate state is not called gas, but plasma. We have already discussed plasma flows in section 7.2. They flow through the entire cosmos as Birkeland streams and can take on huge dimensions without being visible to us. As long as no recombination takes place in the plasma, this is called the dark mode of the plasma. The current density is very low. The glow mode is reached when the current density is so high that the recombination begins and the electrons release their energy to dive into the electron orbits of the atoms. Finally, we still have the arc mode when the current density will be so high that it comes to a sudden discharge of the plasma. The idea that the sun is powered electrically derives from Ralph E. Juergens [7.46] and Hannes Alfvén [7.50] who both designed two different models. While Alfvén's model was not further developed, Juergens' idea found a fertile ground, as it could explain some important phenomena.

Juergens concluded from the structure of the solar atmosphere,

> *» ...that the solar energy must be fed not from the inside, but from the outside, and the energy delivery mechanism is an electrical discharge.«*

His hypothesis of an electrical discharge associates the sun with an anode and the photosphere with the anode glow light. He recognized the seen granules on suns surface as anode tufts, an indication of a self-organizing open system. Between the stars we have a positive column, or a dark mode and the cold space is the virtual cathode. The current always flows into the sun via the anode tufts, while the current flow direction in Alfvén's model depends on the solar cycle. In 1968, from the observations of the Fraunhofer lines it was inferred that both neutral atoms and positive ions would have to drift between the anode tufts [7.52]. That would be a way to build up double layers.

Juergen's Anode Sun model can explain why temperature inversion in sun's atmosphere exists: It can explain the existence of the corona, the solar wind, and why the spicules in the chromosphere are, what they do and what they exist for. It may also explain why the anode tufts die away. This model was significantly developed by Don Scott. His book "The electric sky" [7.53] was published in 2006. Don Scott dealt on this basis with the questions of why there are two varieties of solar wind? Which electric process can bring the solar wind to a standstill over two days and what can accelerate the solar wind? Can sunspots be integrated into the model? Scott thinks that the fast solar wind comes from the photosphere clumps and the slow wind comes from the sunspots around the solar equator. With his transistor solar model, he can explain the different speeds and densities of slow and fast solar wind. The solar wind has a mean proton density of about 6 particles per cm³ with a fluctuation of 3 particles per cm³ by a calm sun, as can be seen on http://spaceweather.com. The average speed of the solar wind is about 350 km/s with a fluctuation range of about 50 km/s. However, the electric solar model had the disadvantage compared to the classic solar model, it could explain a large part of the phenomena observed on the sun, but it could not explain the nuclear fusion.

The electric solar model could not explain the nuclear fusion.

Scott's electric anode sun model sees the sun as an energy consumer. It does not answer why the Sun stays at a positive potential and how element fusion occurs and what causes the potential difference and why the sun should bee an energy sink. The fusion also provides energy. But is it sufficient to satisfy the amount of energy need emitted if no appreciable energy is released from within when there is no neutrino current?

Let us remember what we covered in section 6.5. The sun is then to be understood as an open system. It absorbs mass and energy from the environment and emits entropy as radiation and solar wind. However, we still know too little about it, only so much that there are dissipative structures in the form of photosphere tufts on the sun. This entitles us to assume that the entered entropy is less than the given entropy. The external entropy is therefore less than zero. As a result, their amount is larger than the internal entropy in order for dissipative structures to grow in the form of photosphere tufts.

Tabelle LXVI. Wahrscheinlichkeit der Kernreaktionen bei $2 \cdot 10^7$ Grad

Reaktion	Q	P (sec^{-1})	mittlere Lebensdauer
$H^1 + H^1 = D^2 + \beta^+$	1,53	$8,5 \cdot 10^{-21}$	$1,2 \cdot 10^{11}$ Jahre
$D^2 + H^1 = He^3$	5,9	$1,3 \cdot 10^{-2}$	2 Sekunden
$H^3 + H^1 = He^4$	21,3	$1,7 \cdot 10^{-1}$	0,2 Sekunden
$Li^6 + H^1 = He^4 + He^3$	4,1	$7 \cdot 10^{-3}$	6 Tage
$Li^7 + H^1 = 2He^4$	18,6	$6 \cdot 10^{-4}$	1 Minute
$Be^9 + H^1 = Li^6 + He^4$	2,4	$6 \cdot 10^{-13}$	2000 Jahre
$B^{11} + H^1 = 3He^4$	9,4	$1,2 \cdot 10^{-7}$	3 Tage
$C^{13} + H^1 = N^{14}$	8,2	$2 \cdot 10^{-14}$	$5 \cdot 10^4$ Jahre
$C^{12} + H^1 = N^{13}$	2,0	$4 \cdot 10^{-16}$	$2,5 \cdot 10^6$ Jahre
$N^{14} + H^1 = O^{15}$	7,8	$2 \cdot 10^{-17}$	$5 \cdot 10^7$ Jahre
$N^{15} + H^1 = C^{12} + He^4$	5,2	$5 \cdot 10^{-13}$	2000 Jahre
$O^{16} + H^1 = F^{17}$	0,5	$8 \cdot 10^{-22}$	10^{12} Jahre
$Mg^{26} + H^1 = Al^{27}$	8,0	10^{-26}	10^{17} Jahre
$He^3 + He^4 = Be^7$	1,6	$3 \cdot 10^{-17}$	$3 \cdot 10^7$ Jahre
$Be^7 + He^4 = C^{11}$	8,0	$3 \cdot 10^{-30}$	$3 \cdot 10^{20}$ Jahre

Figure 7.22: Nuclear reactions at 20 million degrees
Reference: Schpolski Atomphysik Bd.II

The observed temperatures, even in the solar corona, do not seem to be sufficient for thermonuclear fusion. E. W. Schpolski [7.61] indicates a temperature 10 times higher. (compare to Figure 7.22) Nevertheless, there must be a fusion of elements in the solar atmosphere. This follows from the evolution of the stars, as reflected in the Hertzsprung Russell diagram. More than 90% of the stars follow the main strand. In an electric solar model, we have no thermonuclear reactions in the interior of the Sun, just as we have no open combustion inside a glowing ash pile.

Candidates for fusion centers are the solar flares, electromagnetic eruptions that rise into the corona and ionized atoms from the surface and deeper layers. Solar flares (Figure 7.23) are the transition from the glow discharge to an arc discharge [7.62].

This happens when the energy stored in twisted magnetic fields (usually via sunspots) is suddenly released. They produce a burst of radiation across the electromagnetic spectrum, from radio waves to X-rays and gamma rays. Solar flares extend to the corona. Within such a solar flare, the temperature typically reaches 20 million Kelvin and can reach up to 100 million Kelvin [7.63]. The magnetic fields in the arcs may then be strong enough, for the fusion of the 'naked' cores to generate heavier atomic nucleus's. The necessary protons are always present in interstellar space as spectra of galaxies show. They are part of the cosmic plasma. Another candidate for a fusion center could be the lower corona, where, according to Scott, the high-swirled atoms are deprived of their electrons, fusion with the protons of interstellar space and then sink back to the suns surface.

Figure 7.23: strong Sun eruptions
Reference: SDO Mission NASA

An indication of fusion is the X-ray radiation. X-rays can arise in two ways. The first type is when an electron beam strikes an obstacle and decelerates sharply. This type is also called Bremsstrahlung. The second type of X-ray formation occurs in the inner shells of atomic shells of higher atomic number chemical elements.

This requires that an electron from one of the atomic shells closest to the nucleus be captured by the atomic nucleus beforehand. This triggers a cascade of further electromagnetic radiation until the atomic shell is in equilibrium again. This happens during the merger. The fusion process produces a positive charge surplus, as we saw in Chapter 5. This is compensated by the capture of electrons from the deepest layer of the atomic shell. The consequence of this is a cascade of emission jumps of the electrons from the overlying shells. These emission jumps produce the short-wave X-radiation. As the **S**olar **D**ynamics **O**bservatory (SDO) images in Table 7.7 show, the likelihood of nuclear fusion centers is not evenly distributed across the solar surface. Only in the brightest centers can you suppose fusion processes according to Schpolski.

Looking at the charge ratio of hydrogen and calcium for neutral atoms, it changes from 1 to 1.87 till 2.0 (see Table 5.1). This means that with increasing duration of fusion, a chronic electron deficiency occurs, which must be compensated by inflow of electrons. This in turn means that near the sun the hydrogen loses its electrons and the remaining protons are rejected as solar wind. The SDO also provides daily measurements of the proton current of the solar wind.

Now we know where nuclear fusion takes place on the sun, but we still don't know how it comes about. Don Scott comes to our aid again. He managed to identify the butterfly nebula as a Birkeland stream. There you can clearly see that the electricity is constricted in the middle. We already got to know the force that causes this in section 7.2 The Birkeland Current.

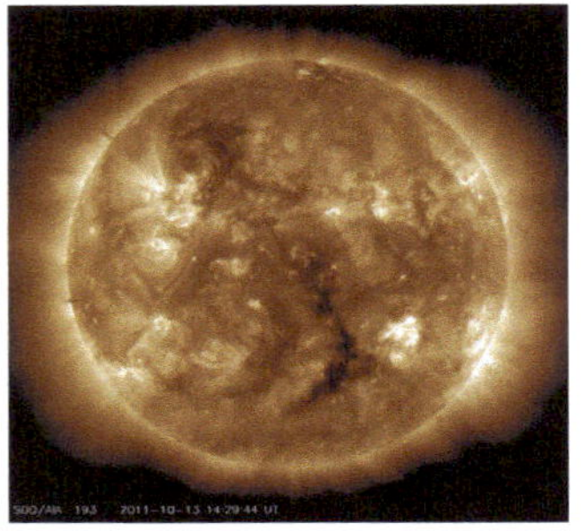

X-ray image of the sun from 13.10.2011 without solar wind. The corona is distributed fairly evenly over the sun's surface. North and south of the equator, category C2 flares are visible.

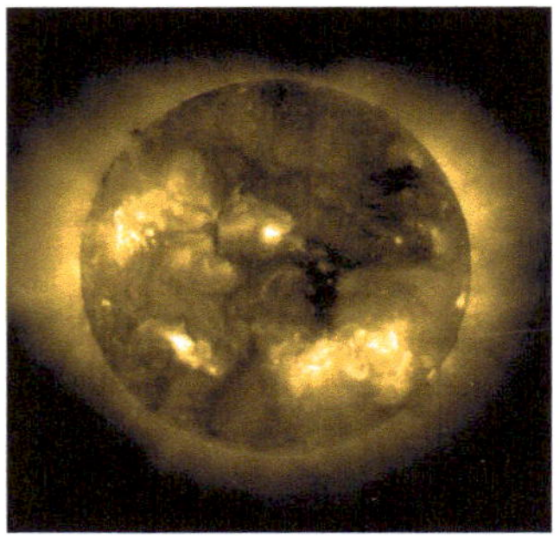

X-ray picture of the sun from 1.09.2001. Solar wind 1P / cm³. The corona is still distributed over the entire solar surface. But there are large flares of category M and X visible.

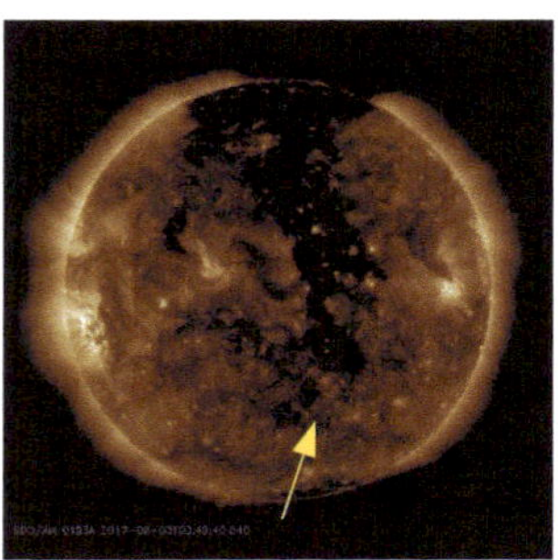

X-ray picture of the sun from 11.12.2017
Solar wind 40.7 P / cm³. The corona has significant holes at the poles and the northern hole extends far into the southern hemisphere. Only weak flares of category A9 and B6 are visible. There is a small sunspot in the left quarter on the equator.

Table 7.7: Images from the Solar Dynamics Observatory (SDO) SDO / AIA 193; to 20 million Kelvin in the white spots corresponds to 1.7keV Reference: SDO Mission NASA

This force can obviously be used to create a nuclear fusion. You have to think of it like this: If the particles all roar at high speed, like on a freeway, there are no collisions. However, as soon as the current narrows, there is a jam and the particles crash into each other. A new star lights up in the center of the butterfly nebula.

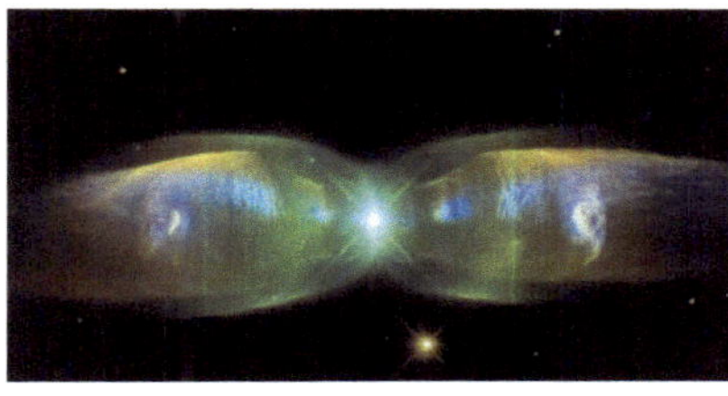

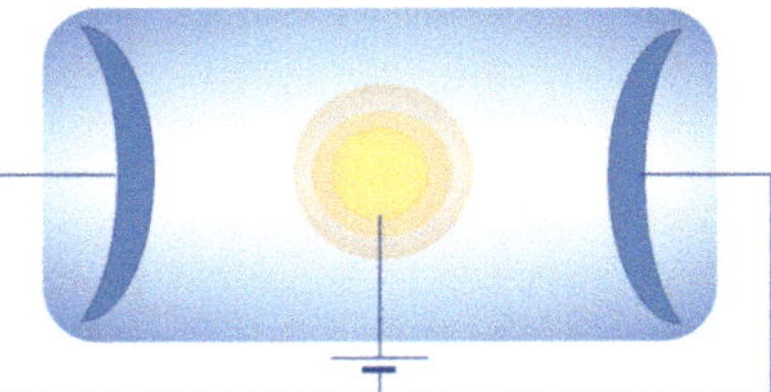

Table 7.8: The butterfly nebula M2-9 is the model for the technical fusion experiment SAFIRE

In 2013, Montgomery Childs and a small team began to replicate Scott's solar model in the laboratory by observing the self-assembly of a plasma consisting of protons and nitrogen ions in a vacuum chamber. To generate the plasma, two opposing large plate-shaped electrodes are arranged in the vacuum chamber, between which a small spherical copper anode is situated, which represents the solar model.

The project was named SAFIRE. The name SAFIRE stands for **S**tellar **A**tmospheric **F**unction **I**n **R**egulation **E**xperiment. For this experiment, a lot of parameters, such as gas composition, pressure and temperature, voltage and current have to be regulated. Likewise, the metrology of these parameters must closely monitor and controlled. The model has shown that it is possible to keep stable high-energy plasmas. Passing through the current-voltage characteristic resulted in several stable levels, in which tufts on the anode first appeared, later double layers and finally a very hot plasma. These double layers are strong self-organizing electromagnetic fields, as discussed in section 7.7, which are responsible for stabilization. They vary in intensity and number by a number of interacting factors as well as voltage and current. In this case, operating voltages in a range of 300 to 400 V and a current of 1.5 to 2 amps and a chamber pressure of 20 Torr were used. The work was done with a nitrogen-hydrogen mixture, as it is found interstellary in the spectra of the galaxies.

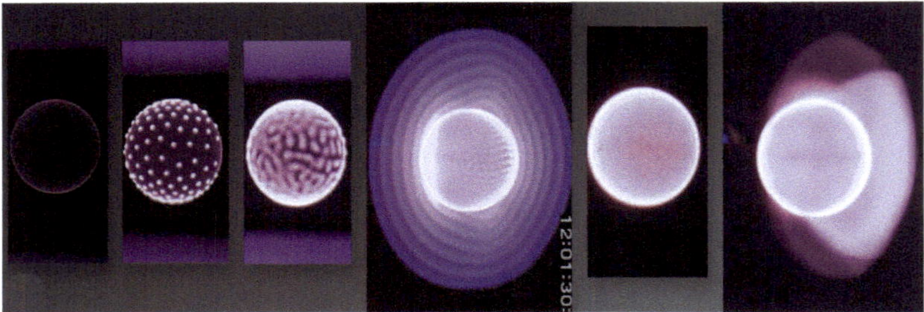

Figure 7.24: Anode images in SAFIRE
Reference:https://www.youtube.com/watch?v=DTaXfbvGf8E

This work had to be carried out very carefully in small steps, as there was always the risk that the entire system could be destroyed as a result of an uncontrolled lightning discharge. With the sun model of the SAFIRE project, many of the assumptions of Scott's anode model were confirmed. It was also observed that the temperature in the immediate vicinity of the anode first decreased and then rose again sharply with increasing distance [7.53].

The stronger the current, the hotter the plasma and the anode became and there was a risk of anode starting to melt. At the predicted power limit, however, only 7% of the power actually used was reached and the experiment had to be stopped. Where did all the energy come from? The anode had changed a bit.

After the experiment, the anode was examined under the electron microscope for possible deposits. In fact, smallest droplets of various chemical elements from the plasma condensed on the anode surface, which were visible under the electron microscope, chemical elements that were not previously present in the chamber.

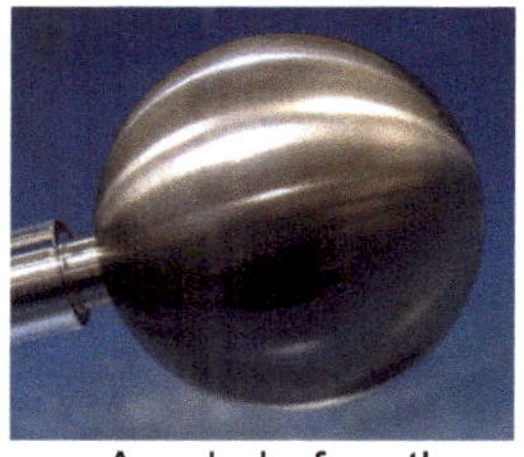
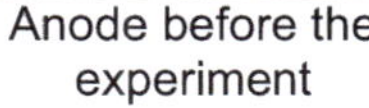
Anode before the experiment

Just before melting

The anode after that

Table 7.9 :- Reference: https://www.youtube.com/watch?v=DTaXfbvGf8E

Deposits highlighted in white on the anode

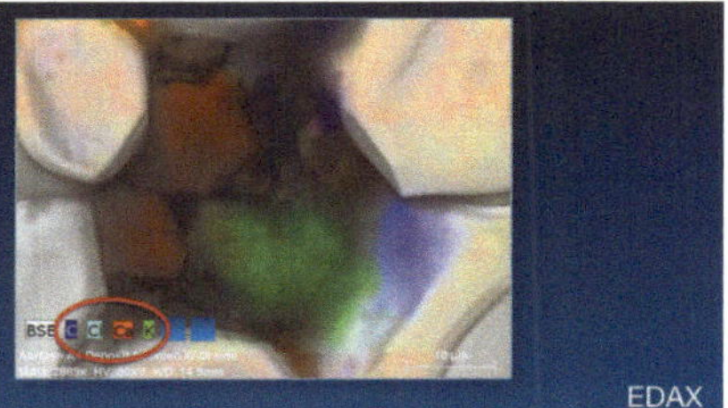

Combined chemical elements highlighted in color

Table 7.10 – Reference: https://www.youtube.com/watch?v=DTaXfbvGf8E

This finding supports the droplet model of the atomic nucleus, as described in section 5.4. Apparently, the structure of the atom begins with the building blocks deuteron and triteron, which fuse from the hydrogen molecule with the release of an electron with one electron each within the plasma to form a deuteron or an additional proton to form a triteron. To do this, the molecular hydrogen must be ionized in an electric field and robbed of one of its shell electrons at the anode. Two protons and one shell electron remain. But this electron now gets between the two protons and so they merge into a deuteron by giving off energy. If both electrons are lost at the anode, the protons are sent to the cathode as a solar wind or captured by a deuteron. This process corresponds exactly to the first case of Prigogine's theorem about entropy in open systems. The system removes more entropy than is generated internally. This is exactly the effect that provides the energy that we want to use technically in nuclear fusion.

These building blocks made up of a core electron and two protons or three protons to a lesser extent represent elementary magnets that contract into larger units. They could be detected by mass spectrometry in the SAFIRE chamber.

All elements were found that were also identified in the Fraunhofer lines of the sun spectrum. All elements were found that were also identified in the Fraunhofer lines of the sun spectrum.

7.10 Is the Terrestrial Weather Electric?

The sun is our energy source and our immediate reference system. The interaction of the sun with the earth, the surface of which is covered by about 2/3 of water, creates a weather situation in a certain area at a certain time which we understand essentially as a thermal consequence of solar radiation and which we register as high pressure and low pressure areas. But also the solar wind and cosmic radiation from the depths of the Milky Way, so positively charged particles are constantly pouring towards the earth. Although most of these particles are intercepted by the earth's magnetic field, but because of the good correlation between cosmic rays and the solar wind index of the last thirty years, additional information on the effects of the solar winds on the terrestrial weather for the past 150 years, is possible by extrapolation, Horst Borchert claims [7.55]. He observed a high level of radiation intensity in times of low sunspot activity and vice versa. A theory put forward by Henrik Svensmark states that changes in the magnetic field of the sun have an influence on the cosmic rays that reach Earth [7.56]. Not only the aerosols introduced by humans, but also ions have the ability to condense water vapor in the troposphere. The latter ability is exploited in nebulous chambers for the detection of radioactive radiation. However, the earthly and solar influences are very controversial under political aspects.

In the meantime, CO_2 has been identified as the main culprit for global warming. However, the clouds are not even considered in the climate models. Because of their size and quantity compared to CO_2 molecules, aerosols in connection with clouds are likely to have a far stronger effect on the back reflection of the heat radiation from the earth than one ever thought. However, it is currently believed that hydro-sols are more likely to be responsible for cooling. But back to the influence of the electric sun.

Our aurora borealis is the visible part of the birch stream flowing through the earth. This has to be seen in the stratosphere as a high current. Our stratosphere is positively charged due to the constant solar wind. We have learned that a force-free current can be described by a Bessel function. The flow profile then follows a Bessel function of the 2nd type according to Scott. The typical characteristic of such a flow is the counter rotation, as can be seen in the following figure 7.25.

Such a pattern of movement has also been observed on other planets. Even with galaxies counter rotation, was observed in their rotational movement [7.57] as we already mentioned in section 7.4.3. However, it is striking that the strong power is formed facing away on the sun side, namely the polar night, while no vortex has formed around the polar region in the southern hemisphere. On the ground, the atmosphere is carried away by the rotation of the earth. The influence of the Birkeland current and the entrainment effect of the earth's rotation causes a vertical and horizontal swirling of our atmosphere. This constellation causes a predominantly northwestern current in winter for Central Europe.

Now let's look at the question raised in section 6.2: Why do we have clouds in the sky? Clouds consist of the smallest droplets of water, at which the light is scattered and which are clearly heavier than air. According to gravity, the clouds should not be able to float in the sky. But they do not fall from the sky. They sometimes rise to heights where the droplets freeze to snow and ice. Conventional meteorologists observe the air pressure and then determine the weather outlook, but they can not say whether rain will fall from the dark cloud or the clouds continue to move. If they then discharge themselves, sometimes several liters of water fall in a very short time on the square meter. That's several million liters per square kilometer, which can lead to huge damage in the affected areas.

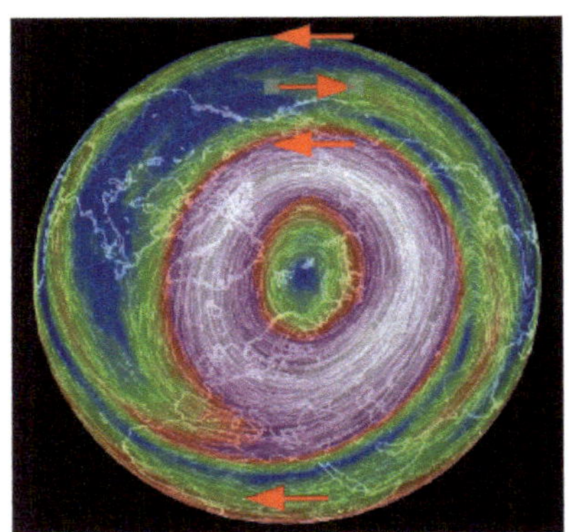

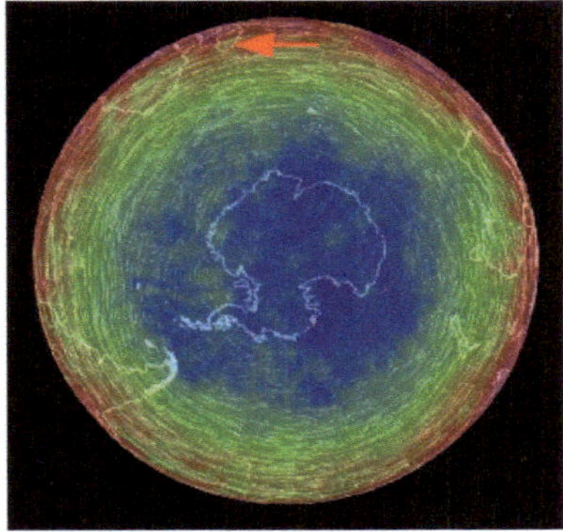

Figure 7.25: Altitude flow at a height of 26 km in January 2020
Reference https://earth.nullschool.net/

From this we conclude that a rain cloud can contain many thousands of tons of water. Clouds have no aerodynamic properties that allow flying and they resist the gravitation after all. We say it's the thermals, the warm air that blows up the clouds. But warm air does not generate electricity yet.

Why does a single cloud often form over the sea near the coast? Would not one have to expect a uniform fog layer? Now you can argue that there are strong updrafts in the game. That's right, but what drives these air masses? The temperature differences and thus the differences in the air? This does not explain the grains of ice falling from storm clouds and it does not explain the different cloud heights. We only know only one force that can carry such a burden. This is the force that acts between equal electrical charges. The air flow has to carry electrons upwards and the earth is charged regionally positively. But meteorologists know nothing about it, at least the electric charge does not matter in their weather forecasts. The reason for the behavior of the clouds lies in the special properties of the water.

The explanation is a bit more complicated than you might expect. We know, that our troposphere over an altitude of 60 km, in its under part the weather is taking place, is enveloped by an ionosphere that reflects radio waves. This provides positive charge down and it is constantly supplied by the solar wind and the cosmic radiation with new charge carriers. As a result, their charge concentration changes with the day-night rhythm. Our earth has a negative charge due to the natural radioactive decay of the earthly elements. This results in a terrestrial mean electric field of about 100 V/m height. This can locally increase to lightning discharges up to over 300,000 V/m by shifting of charges. Water in the atmosphere is responsible for the charge exchange between the cosmos and the earth's surface. For this, the water droplets in the clouds have to transport negative cargo. This creates the necessary force to keep the droplets in the sky. At the same time, the droplets must also have dipole properties in order to be able to form larger units, and on other hand they store protons from the high atmosphere. So they grow up on

Figure 7.26: The Cloud

their way until they have neutralized enough tc fall down. But why do water droplets carry a negative charge?

To understand this, we need to take a closer look at water, as the stuff most commonly encountered on the earth's surface. We guess that we know the properties of the water. It meet us every day, but we do not waste any thoughts on its peculiarities. We all know the phase diagram of the water and we may have wondered that it has the highest density at 4 degrees Celsius and that ice floats on the water, or that snow crystals have a hexagonal shape, maybe also that there are isolated clouds in the sky, without thinking about it.

We are used to see water as a collection of individual molecules that are completely independent of each other. That is most wrong. Especially in the transition between two different phases, we observe a behavior that the molecules force into a self-chosen order, which causes a swarm behavior or a structural formation. In section 6.5 we learned that texturing is always associated with entropy removal. The structure of the water at its phase transitions is a lot more complicated than we would expect. Let us consider a very simple example of self-organization, which was pointed out in 1990 by Herman Haken [7.58]. Take a pot of water and heat it gently and watch what happens. At the hot bottom of the pot, vapor bubbles begin to form, which rise to the surface and produce a well-ordered flow pattern, a dissipative structure. These vapor bubbles occur regularly at always the same places. But Boltzmann's thermodynamics assume a very irregular movement of the molecules. In fact, it is not individual molecules that are released into the air, but the exiting water molecules are or-

Figure 7.28: Thermal image of the Cell structure of hot water Reference: G. Pollack

ganized in small negatively charged bubbles, as Gerald Pollack has noted [7.59].

He found that water is arranged in the vicinity of a hydrophilic (water-loving) phase in a very specific structure which he calls the 4th phase of water. It forms a double layer. A double layer is a battery, as we have already learned. When water is exposed to an electric field, it dissociates into H+ and OH-, H2O and OH- give H3O2- and H+. The protons form the bridges between them. (see Figure 7.29) These are the basic building blocks for a hexagonal structure of water. The corresponding negative sides of the structure then collect in the immediate vicinity of the hydrophilic carrier layer, as is also referred to in Figure 7.20 as the Helmholtz layer. It could be seen as a liquid crystal in the immediate vicinity of the hydrophilic phase.

In conventional physics, such a structure is rejected because there is represented the dogma of the quasi-neutrality of matter. Neutrality means, all forces are in equilibrium. However, this in turn means that nature is dead. But we profess a living nature far from the thermodynamic equilibrium.

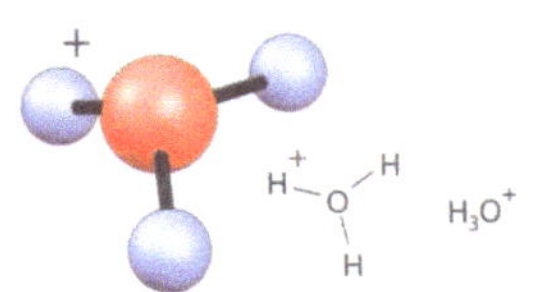

Figure:7.29 Hydronium ion - Reference: G. Pollack

The hexagonal ordering of the water molecules results in a honeycomb-like planar structure of many negatively charged leaves that are stacked on top of each other like slightly shifted book pages without compensating the charge. This quasi-crystalline negatively charged structure displaces the ordinary water from the surfaces. Pollack called this zone *Exclusion Zone* abbreviated *EZ* or simply *EZ-water*, which exclusion contaminants with emulsions in it. Due to their negative charge, a large amount of protons accumulates around the EZ-water these can not penetrate into the dense crystal structure. Thus, this charge separation is retained as in a conventional battery. Ordinary water remains further away from this layer. Pollack speaks of bulk water.

The fact that ice is lighter than water is explained by Pollack with the honeycomb structure of the EZ-water, where protons are stored in, causing the transition to ice and reducing the density of water [7.60]. The hexagonal shape of the snow crystals a proof of that and they should be negatively charged.

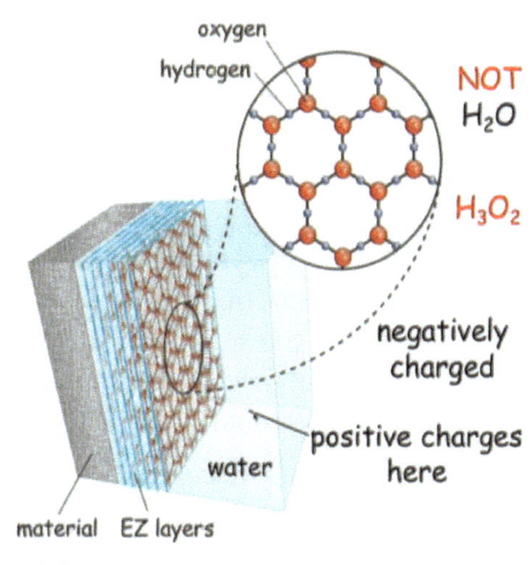

Figure 7.30: Water battery Reference: G. Pollack

If you now see the rising vapor from a hot pot of water against a dark background, you do not have individual water molecules in front of you, but accumulations of vapor bubbles of at least one micrometer in diameter, because light can only refracted by objects which are slightly larger than the wavelength of visible light. This corresponds to about one billion molecules emerging discontinuously in the dressing. The emerging also occurs only on the patterns appearing dark in Figure 7.31 on the surface of the water. Pollack identified these patterns as structured

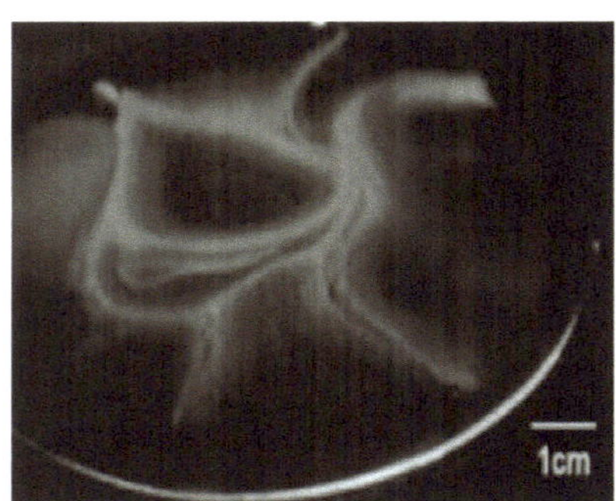

Figure 7.30: Thermal image of the steam structure over a hot water surface
Reference: G. Pollak

negatively charged EZ-water vesicles (droplets and bubbles). To understand the clouds, it is now important that the ordinary water droplet or bubble is covered by a membrane made of EZ-water.

Thus, the negatively charged vapor rises tubular upwards and it formed vapor vesicle on vapor vesicle, when sufficient charges have accumulated, which are surrounded by a layer EZ-water. The smaller these vesicles are, the larger the ratio of surface area to volume, and the more negative the vesicles are charged. But the vesicles stick together and form clusters, which should be responsible for the positive charge inside the clusters. At a certain height above the water, the vapor structure disappears as a result of the vapor pressure sinking. The structure dissolves and the vapor vesicles mix with the air and the aerosols (fine dust), which then become hydro-sols and penetrate into the clouds. We recognize these hydro-sols by the dark color of the clouds in reflected light. Its effect on the weather has not yet been investigated. But they are likely to be responsible for the increase in heavy rainfall. Perhaps the formation of snow due to the slight warming of the cloud is prevented.

Now, how does it come to cloud formation? On the one hand we have the negatively charged vesicles and on the other hand the positive protons in the troposphere. There where the two types of charge get together and contract, cloud formation occurs, provided the dew point is reached and the bubbles condense to droplets. This requires a cooling of the surrounding air, which is guaranteed only at a corresponding height above the earth. In addition, remember that the clouds move in an electric capacitor field whose negative electrode is the earth, caused by the EZ water at the sea surface and in the vegetation and whose positive electrode represents the ionosphere. The earth is negatively charged and the clouds are negatively charged. This results in a repulsion. Depending on the amount of negative charge, the clouds are then either higher in the sky or lower.

As a result of the earth's rotation and the electrostatic attachment of the atmosphere to the negatively charged earth's surface, we have a relatively calm atmosphere. As soon as the local charge conditions change, this expresses in wind itself. Thunderstorms are always accompanied by larger potential differences, which are indicated by lightning discharges. They are also usually accompanied by storms. Rain then arises from the fact that under the negatively charged cloud a positive charge accumulates in the soil, which pulls the rain out of the cloud.

The weather lighting in the clouds can be explained by the charge balance due to the change in the charge ratios between the surface and volume of bubbles during the growth of bubbles to raindrops. Large charge differences between earth and cloud or cloud and stratosphere then lead to various types of electrical discharges, triggered by cosmic radiation that build up the discharge channel. These phenomena, summed up under the name Sprites, were observed in particular on the night side of the Earth by the **I**nternational **S**pace **S**tation (ISS). The cloud cover connected to the Earth's radiant belt is likely to have a certain buffering function with respect to electrical discharges by preventing electrical discharges from space from directly penetrating the earth.

8 More Cosmic Puzzles

8.1 Planetary Crater Surfaces

If we look at the images of the earth's moon, from Mars, from Mercury and from comets, we notice everywhere the many craters that standard astronomy explains with the impact of meteorites. All of these celestial bodies either have no atmospheres or, if they do, they do not have clouds of water. For a number of these craters, the meteorite hypothesis may apply. But there are also craters with features that make one doubt on the meteorite hypothesis. In particular, it was noticeable that many craters are circular, which can happen only in a vertical impact. Other craters have a cone inside, and others have smaller craters on the crater edges. The sheer size of some craters makes us wonder that the celestial bodies were not broken by the impact. Then you can find scars, deeply dug valleys, without seeing where the excavated material has remained. Running water would have relocated the material only.

We know on earth the electrical erosion for the purpose of material removal. Here, the material is evaporated and the material removal can be considerable in the case of arc discharges. There must have been a very dramatic time in the evolution of the planetary system that has left such traces of electro-erosion. Considering that the planets with the sun are in a Birkeland current and can form double layers through dust and gases, an explanation for this type of the scar formation is close. If negatively charged celestial bodies come close to positively charged celestial bodies, they will discharge material from the positively charged celestial bodies by means of discharge arcs until the equi-potential bonding is established. A direct impact of meteorites is not necessary at all. An approximation is enough for an electrical arc discharge. The knocked-out material should exert an impulse on the rest of the celestial body, which, like by a rocket, should lead to a change in the orbit. Also, the highly eccentric orbit of comets could be explained by assuming, that comets represent fragments of electrically excavated planetary material.

These few examples show that the acceptance of electric plasma currents makes cosmic phenomena much easier to understand than the so-called standard cosmology with conventional prescriptions of gravity succeeds. If you want to get deeper into the matter of plasma cosmology, you will find a lot of material under the Thunderbolts Project^{TM}. There is a menu item **you-tube** with many video clips on the subject. There is a menu item space news with many video clips on the subject. However, while scientists like David Talbott [8.05] of the Thunderbolts project believe that there must have been an electrically very active phase in our planetary system in prehistoric times due to traditions from ancient myths, I am quite skeptical about this. The evolution of galaxies shows that once there was a phase of high electrical activity in each galaxy. But this period must have been very far in the past and is to look for in the initial phase of the formation of the solar system .

There have been attempts to relate the circulation of the sun around the milky way to the history of the earth. There was a global catastrophe at the end of each period in Earth's history that was crucial to life on Earth, but the timings have a different period than the Galactic year. The Earth History covers about 16 galactic years. With the entry into the Holocene began 12,000 years ago, the warming of the earth. It is unknown, what could be the cause. According to the latest findings, it is likely that it have to do with a changed activity of the sun and deep into the atmosphere penetrated solar storms. But these are speculations without hard evidence.

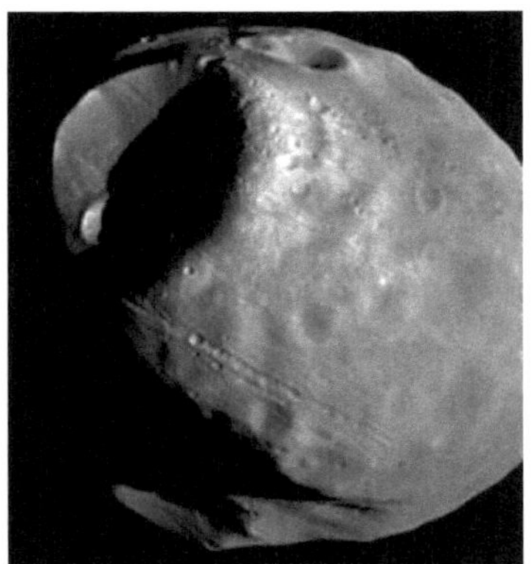

Figure 8.1: Marsmond Phobos Reference:Viking-Project, JPL, NASA;

Earth-bound local catastrophes are likely to have burned far more into the memory of humanity than changes in the depths of the planetary system, especially since in antiquity the idea of the depth of the cosmos was hardly developed. In contrast, large sun-flares may have already been noticed and their effects on the earth perceived. But such dramatic changes, as Velikovski describes in his book *Welten im Zusammenstoß*, surely would not have survived humanity and with it the whole biosphere. Why should one trust the myths of antiquity more than modern myths, such as the theory of relativity?

We only can make assumptions about the early days of Earth's history based on the interpretation of a few conserved tracks. But also it could have looked different. From the comparison of galaxies we can conclude that there must have been an electrically very active phase in our Milky Way. The traces of it have been preserved in the craters on the planets and moons without a biosphere. But we are not aware of any changes in the planetary arrangement within the solar system. These would have been far more dramatic for our planet than the traces of the earth's history testify. An idea of this is perhaps the crater on the Mars Moon Phobos.

8.2 White Dwarfs

Another mystery is the so-called *White dwarfs* who are outside the main line of the Hertzsprung-Russell diagram who's luminosity is two orders of magnitude lower than that of the sun. According to WIKIPEDIA, White dwarfs are said to represent the final stage in the development of a relatively low-mass star whose nuclear energy supply has come to a halt. They would be the hot nuclei of red giants that would remain if they repelled their outer shell. The prerequisite for this is that the residual mass remains below a threshold value of 1.44 solar masses, the so-called Chandrasekhar limit4) assuming that White dwarfs are electrically neutral. Otherwise, after a supernova outburst, a neutron star or (with a nuclear mass of more than 2½ solar masses) even a black hole would arise , which is physically not explainable. Free neutrons disintegrate within minutes and there is no space curvature, as we have already discussed.

The nearest White dwarf is Sirius B, the tiny companion of Sirius, the brightest star in the night sky. The 8.5 light-years away, very hot Sirius is 2 times larger and 22 times brighter than the sun. Sirius B is been estimated due to its brightness only earth size, but to 98% of the solar mass and 2% of their luminosity. It is the most studied star of this type. A teaspoon full of its mass is supposed to weigh over 5 tons on earth. According to our physical understanding, the atomic shells on a white dwarf should be much smaller. This should result in completely unknown optical spectra, since they would be shifted far into the blue, but Table 8.1 shows a strikingly strong absorption line of the hydrogen at the known location against a completely normal thermal background spectrum.

In 1917 Adriaan van Maanen discovered the so-called Van Maanen's star. It is an isolated White dwarf at a distance of 13.9 light-years. The White dwarfs are explained today by the theory of degenerate matter: Degenerate matter is said to be widespread in the cosmos.

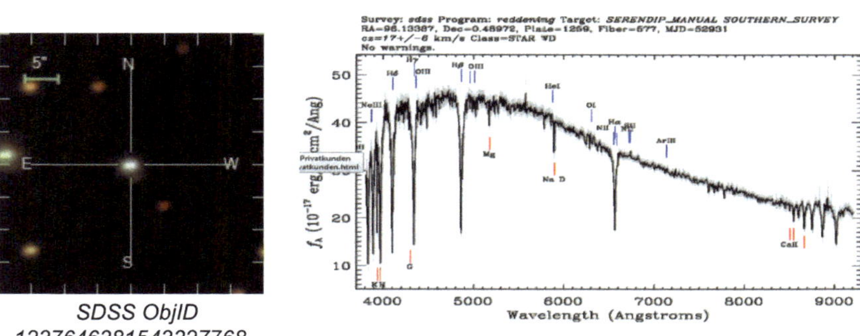

SDSS ObjID
1237646381543327768

Table 8.1 – White dwarf - Reference: http://skyserver.sdss.org/dr14/en/tools/explore

It is estimated that about 10% of all stars are White dwarfs, which were made of degenerate matter (especially oxygen and carbon). In White dwarfs the electrons would have degenerated, in Neutron stars the neutrons. The bigger this hundred-thousand-degree hot stellar mass, the more it would be compressed which contradicts thermodynamics indeed.The degeneration would be caused in the degenerate pressure, which should counteract the gravitation. It's another wild card for ignorance. I found 6 white dwarfs in a sample of 500 stars from the SDSS database. Of these, 5 were isolated stars and one had its maximum radiation in the ultraviolet. The remaining four were very similar. Let's take a look at one of these 4 white dwarfs: If we compare this spectrum in Table 8.1 with that from Table 7.6, the only noticeable difference is that the absorption lines of the hydrogen of the Balmer series are particularly pronounced and the lines are also wider than for stars in the main series. The metallic part is also visible. Nuclear fuel is significantly more available than the G2 star. It is astonishing how the above statements can be derived from these unambiguous findings. Only the higher electron density is evidenced by the line broadening of the hydrogen lines and the surface temperature of around 4800K is shown by the background radiation maximum. With comparable spectra, the size of the star can be inferred from the brightness, but not its mass. The pressure broadening of the hydrogen lines suggests that the star is at the center of a strong pinch force. It may be a young star, like the one standing in the butterfly nebula (Table 7.8), except that you can see directly into the nebula along the Birkeland stream. The rings around the star are too weak to be seen because they are outshone by the star itself.

8.3 Pulsars

A pulsar (means **pulsa**ting **r**adio emission) is claimed to be a fast rotating neutron star. Pulsars were first discovered in 1967 using radio telescopes. We noticed a series of radio signals with a periodicity of 1.3 s from the constellation Vulpecula. About a year later, the astronomer Thomas Gold [8.01] came up with the idea of a fast-rotating neutron star, which is supposed to emit concentrated radiation similar to a rotating light beacon. The symmetry axis of its magnetic field should deviate from the axis of rotation, which is why it should emit synchrotron radiation along the dipole axis. If the earth lies in the radiation field, it regularly receives recurring signals.

Pulsars radiate mainly in the radio frequency range, sometimes down to the X-ray range or only in this area. Of the more than 1700 known sources, only from a few, were observed intensity fluctuations in the visible range. The supposed rotation time of a pulsar without a companion is between 0.01 and 8 seconds. It decreases by about 10-15 seconds per second (that is, it slows down over time), and from that, the experts expect to have a lifespan of about ten million years. In addition, so-called millisecond pulsars (about 5 percent of the pulsars) with a pulse frequency of one to ten milliseconds and a longer life were found. If a star is to rotate at the speed of a dental turbine, something must be fundamentally wrong with the theoretical considerations of the lighthouse principle. At the end of 2016, Wim Hermsen and his team [8.02] reported the discovery of the PSR B1822-09 pulsar, which switches between a radio-loud mode and a radio-silent mode, and the team detected X-ray pulsations, direct following the main radio pulse. This discovery confused all those who believed in the lighthouse theory.

We know that free neutrons convert in hydrogen atoms within a short time. A neutron star with ordinary neutrons would vaporize into a hydrogen cloud in a short time. Therefore, Gold claimed that these neutrons would be degenerate. This argument is too weak. We do not know any degenerate neutrons, let alone their properties. What should neutrons hold together at such high rotational speeds? In addition, Neutron stars should have a density of 10^{15} g/cm³ and a magnetic field of 10^{12} Gauss.

How do we even know about the rotation of these radio sources? The only thing we know is the type of radiation and the clock frequency. These claimed properties contradict elementary laws of electromagnetism. This was also noted by F. Curtis Michel who published a theory of the pulsar magnetosphere with the idea of discharges in an electric double-layer [8.03] because the pulsar flashes have more in common with complex radio signals, that have been caused by lightning in the earth's atmosphere. Lightning can also cause x-rays. Thus, even on these pulsars, lightning in the magnetosphere could produce these observed phenomena. In 1995, Peratt and Healy wrote an article on the radiation properties of pulsar magnetosphere, confirming Michel's assumptions [8.04]. Thus, the lightning discharges from flares to supernovae would be due to one physical phenomenon depending on how long the charge was hold and how much was stored in these double layers of the star.

Closing Words

On the basis of these few astrophysical examples, I tried to outline the prospects for future physics as a homogeneous area of knowledge, an area of knowledge based on nonlinear thermodynamics, the electromagnetic properties of matter, and the principle of self-similarity and asymmetry. Of course, hypotheses have also flowed in here. Science does not work without imagination. But these fantasies must be restrained by reason and they must be based on a foundation of facts. You can not stack up hypotheses by considering a hypothesis as mathematically proven and then on this 'quicksand' founding a new hypothesis. That may work in mathematics. Physics differs from mathematics by practical evidence in reality.

There is an essential difference between mathematics and physics. While the mathematical imagination is unlimited, nature has clear phase boundaries. These phase boundaries make it necessary to always take into account that the installed models are limited in space and time. Only we don't usually know the limits at the time they are drawn up. However, if we did not use inductive reasoning, we could not generate any knowledge. Therefore, physics must provide practical evidence that the model chosen is the only explanation for an observed phenomenon. This was the reason why physics failed in the last century because it lost sight of its social mission.

The social mission of physics is to describe real nature and not to be the maid of theology. Not thinking influences being, but being influences thinking. During the Age of Enlightenment, humans emancipated themselves from Homo philis to Homo sapiens. In the last century, at least in the field of physics, the development has gone backwards. This reverse gear was accompanied by an excessive arrogance towards nature. Dialogue with nature became a monologue. We are not the masters of nature, who can exploit it mercilessly, but only the bacillus of their illness, an illness that causes our planet sink into garbage within a century due to numbers in bank accounts. In times of a worldwide pandemic, it looks as if nature is finally resisting, in order to preserve its creation, or more precisely, its evolution.

If physics wants to take its social mission seriously again, it should make a contribution to the entry into the use of solar energy based on nuclear fusion, as outlined in this book, and take care of closed, sustainable economic cycles that minimize the loss of entropy. I see that as a task for future generations. Our and our children's generation have focused on growth and competition. Only constant growth with limited resources is not possible, and chess is a good parable for the competition. The profit is bought with the loss of resources.

Epilogue

As I wrote this book, I kept asking one question: How can it be that highly skilled people are spreading such nonsense in physics? Have these scientists lost track of reality with their growing specialization? Oddly enough, in the past century, one-way thinking without alternatives could spread in specialist areas of physics, which is also reflected in German politics under She-Chancellor Angela Merkel, a she-physicist. There is no public scientific discussion of possible solutions. The Peer-Review System, once conceived as a quality assurance of scientific papers, has become a means of discarding new ideas. In particular, from the data obtained from space since the middle of the 20th century gives a completely new view of the entire physics. Without the internet, these ideas would be lost.

The longer I spent researching for this book, the more I got the idea that there had to be a plan and a network behind such a serious failure in the field of cosmology. A free science would produce more strife. Who has a global network available to manipulate people? My suspicion fell on the Catholic Church. Once Galileo Galilee said, "*The Bible shows the way to go to Heaven, but not how Heaven works,*" but the Church believes that it must control science, that was my working hypothesis when I was working on the manuscript of this book.

As I worked on the correction of the manuscript, I received the confirmation of my working hypothesis by K. Gebler, a mathematics and physics teacher. He referred me to the encyclical Pascendi Dominici gregis of Pope Pius X. in 1907. There is written in section 46,

> *»For in the vast and varied abundance of studies opening before the mind desirous of truth, everybody knows how the old maxim describes theology as so far in front of all others that every science and art should serve it and be to it as handmaidens (Leo XIII., Lett. ap. In Magna, Dec. 10, 1889). We will add that We deem worthy of praise those who with full respect for tra-*

dition, the Holy Fathers, and the ecclesiastical magisterium, undertake, with well-balanced judgment and guided by Catholic principles«

Clear instructions are given in this encyclical on how to counter modernity, how to control the media, and how to become active in science. Furthermore, all aspirations that serve this purpose should be priced. Although Pius X was unable to implement its backward-looking course, because 'modernity' intended at reconciling science and belief, such reconciliation could only be done in favor of faith.

These instructions were apparently very successful mediated by the Pope's soldiers to the mainstream of physicists and their complacency with praise and recognition up to Nobel prizes considered. Nobody seems to care that the logic was damaged. Thus, under the guise of alleged mathematics, a pseudoscience could be bred that poisoned people's brains. The arduously won enlightenment is now recklessly put at risk by physicists for the goodwill of the curia and subverted our educational system with creationist ideas, as it depends in this encyclical. Even the brains of the materialists have been successfully nebulized over several generations. Does not physical abuse always start with mental abuse in all authoritarian systems? The exclusive sovereignty claimed Communist rulers as well as the Pope. Even in the method of enforcing this claim, I could not identify any significant differences between the two camps.

This makes it understandable that the dialogue with nature has increasingly become a monologue of scientists. I have therefore put a focus on the observer and his dialogue. In doing so, I introduced the aspect of observation instead of wanting to unite different viewpoints and perspectives in a theory of everything, like theoreticians dream. I understand mathematics as a tool for describing natural phenomena and, unlike the general view, do not regard them as a source of new discoveries. In doing so, I pointed out mistakes in using this tool.

The basic approach to a truly modern physics is the open system, which becomes an open process considering discrete states. This idea, originating of Prigogine, replaces Einstein's space-time. Within these open systems, cyclic processes take place at phase boundaries under energy and mass transfer, whereby structures can build up and entropy is dissipated.

In our civilized environment many only know the cycle of money. As input parameters we have capital and raw materials and what comes out - growth? No, garbage! - As I write this, on television comes the message that Amazon does not resell returned goods but destroys them because, it's cheaper, than repackaging them. This reduces the life of a product to a minimum in order to grow capital. The capital is used to produce more waste so that it continues to grow until all resources are converted to waste (entropy) at maximum speed. The goal is the maximum profit. Again, the theory of open systems applies. If you think that through, you start to realize that when all the resources of the earth have been used up, capital has lost its value too. For this case, people are promised the departure from the earth and the settlement of outer space. Again a move-out from paradise? Everything would be technically possible. To do that, we would need a space shuttle that provides enough energy to keep a duplicate of an earthly ecosystem running, and we would also need residents of this system who can live on the scarce resources for centuries. Suppose we could build a spaceship that could move between the stars at 300km/s, then the spacemen would travel about 4000 years to the nearest star system, where there may be a biologically active zone. Our culture of written record is about as old and our earth is already on the verge of ecological collapse, caused by our capitalist economy.

We should always keep in mind that we are only guests for our lifetime in this paradise earth, and not owners. »Be fruitful and increase in number; fill the earth and subdue it.« [Genesis 1:28] is literally taken in the new-liberal capitalist economy, and the paradise turns into a hell for the crowd of the subjects in developing countries. For a company like space settlement, we would need another humanity with a different social model, as existing humanity is clearly unable to handle the global problems of overpopulation and ever-increasing environmental degradation. This is caused by greed and the divergent power interests of ruling elites. It is the same mechanism of action that prevents the physical truth from publishing. Who should seriously believe that humanity, which can not even protect its earthly paradise from itself, is able to save a much smaller system from tipping over?

We need other people with a different social model for such a company as space settlement. But if we have this, we do not need to leave our earth anymore. If I was able to anchor this insight to my readers, and more mistrust of authorities who accept no alternative theories and avoid rational explanations, the purpose of the book is fulfilled and perhaps our children will have better insight.

Finally, let me remember to the *Encyclical Letter Laudato si'* of the Holy Father Francis from 2015 for the UN Climate Change Conference in Paris, dealing with environmental problems and recognizing the universe as an open system: »79. In this universe formed of open systems that communicate with each other, we can discover countless forms of relationship and involvement.«

This creates hope for the future. However, the fight against the old demons in the Curia has only just begun and this Pope seems to be quite isolated there, and climate conferences are based solely on CO_2 emissions, a consequence of the energy hunger of our business organization. However, at last it turns about the destruction of our environment through the robber economy of new-capitalist subsistence strategy, which requires constant growth (of capital). Here, a rethinking of the elites of society is urgently needed. We can not escape this trouble by praying, in the hope of recovery.

»There are no supreme saviors. Neither Pope, nor Caesar, nor tribune.
Producers, let us save ourselves, Decree the common salvation
So that the spirit be pulled from its prison.«

But I fear that the nature of humankind is overwhelmed with this activity because, over the last two thousand years, its moral development has not kept up with the development of its technical potential.

Acknowledgments

For the many suggestions and discussions, as well as the review of the German manuscript, I thank Hannes Täger and for the indulgence of my often spiritual absence during the creation phase of the book, For the indulgence of my often absent-mindedness during the writing phase and for the final correction before going to print, I thank my wife, Dr. Marlies Hüfner. I also thank my old teacher Peter Urban, who implanted in me the love of physics and decisively shaped my future life..

To the picture material

For the permission to use illustrations of Prof. Gerald H. Pollack in this book, I thank the authors mentioned. All other imagery is either edited by the author of the book itself, if there is no source remarked, or is shared-free according to the Creative Commons Attribution License (CC-BY) or "Courtesy of NASA / SDO and AIA's science teams, EVE and HMI " or is subject to the public domain or free license of Wikimedia and some other dictionaries.

References

The references were compiled with care. If somebody does not feel heeded, consider that there are nowadays a jumbled number of ideas on the net, and that it's impossible to know every idea in detail, let alone read everything. I refer here to Jean de Climont's book "*The Worldwide List of Alternative Theories and Criticism*" and where represent from more than 7200 scientists born from 1905 onwards.

I have also learned many things in life without having remember for myself, from whom I have this wisdom. Other is based on experience in two so different social systems, without which I would not have come to these insights. What I have processed here, I consider a representative section of the variety of sources. Other authors may refer to other sources and come to different conclusions. Truth is a rating. It is neither absolute nor relative but often controlled by interests of power. You recognize a correct evaluation only in the context with the entire knowledge area. That needs time, more time then we are spending for thinking in our fast living hysterical so called information society often. You only have time if you leave your professional life and you are free of material worries.

1. **Why this Book?**

    1. H. Ratcliffe - *Stephen Hawking Smoked My Socks;* https://www.amazon.com/Stephen-Hawking-Smoked-Socks-astrophysicists/dp/1612641652
    2. O. Gingerich - *Gottes Universum;* http://www.visionjournal.de/node/2121
    3. R. Haumann - *Die Physik des Nichts;* Eigenverlag Viavetum
    4. B. Gaede - *Why God Doesn't Exist*; http://www.youstupidrelativist.com/index3.html
    5. G. O. Müller - *Sammlung von kritischen Stimmen zur Relativität;* http://www.kritik-relativitaetstheorie.de/projekt-go-mueller/

2. **Something Philosophy Pleasing**

1. W..Heisenberg - *Philosophie der Quantenmechanik;* Reclam Universalbibliothek Nr.9948, ISBN 978-3-15009948-3; https://www.amazon.de/Quantentheorie-Philosophie-Reclams-Universal-Bibliothek-Heisenberg/dp/315009948X
2. R. Andersen - *Interview Lawrence Krauss ;* http://www.spektrum.de/news/auch-physiker-sind-philosophen/1353710
3. P. Marmet - *Philosophische Verirrungen in der Quantenmechanik Chapter 4 Was ist Realismus;* http://mugglebibliothek.de/english/marmet1.htm
4. C. A. Ronan - *A shorter Science and Civilization in China;* Bd.I Cambridge 1978 p.170; https://www.amazon.ca/Shorter-Science-Civilisation-China/dp/0521292867
5. I. Kant - *Die Kritik der reinen Vernunft;* http://mugglebibliothek.de/kant.htm
6. A. Schopenhauer - *Die Welt als Wille und Vorstellung;* https://www.amazon.de/Die-Welt-als-Wille-Vorstellung/dp/373720750X
7. T. Chopra - *Die Heiligen Kühe;* Neu Delhi 2000 p. 54; https://www.amazon.de/Die-heiligen-K%C3%BChe-andere-Geschichten/dp/B01FQZPSS2
8. A. Einstein - *Zur Elektrodynamik fester Körper ; http://users.-physik.fu-berlin.de/~kleinert/files/1905_17_891-921.pdf*
9. O. Roy - *Der falsche Krieg ;* https://www.randomhouse.de/webarticle/aid21317.rhd
10. A. Jablonski - *Einführung in die fraktale Geometrie;* http://quadsoft.org/fraktale/#x1-20001
11. J. Becker - *Fibonacci und der Goldene Schnitt;* http://chorgiessen.altervista.org/jab/goldfibo/goldfibo.pdf
12. B. Mandelbrot - *Die fraktale Geometrie der Natur; https://www.amazon.de/Fraktale-Geometrie-Natur-Beno%C3%AEt-Mandelbrot/dp/3764326468#reader_3764326468*

13. J. Bromand:- *Gottesbeweise von Anselm bis Gödel* (*Suhrkamp Taschenbuch Wissenschaft*). Verlag suhrkamp taschenbuch wissenschaft 1946; ISBN 978-3-518-29546-5.; https://www.-suhrkamp.de/download/Blickinsbuch/9783518295465.pdf
14. K. Popper - Logik der Forschung. 11. Edition 2005, 1934 ; https://monoskop.org/images/e/ec/Popper_Karl_Logik_der_Forschung.pdf
15. G. Boole – *The Mathematical Analysis of Logic ;* Cambridge1847; http://www.archive.org/stream/mathematicalanal00booluoft#page/n7/mode/2up
16. Pope Pius X. Enzyklika Pascendi Dominici gregis section 57; http://www.kathpedia.com/index.php/Pascendi_dominici_gregis_(Wortlaut)
17. I. Langmuir - *Pathologische Wissenschaft ;* https://en.wikipedia.org/wiki/Pathological_science
18. W. Kämmerer - *Einführung in mathematische Methoden der Kybernetik;* Akademieverlag Berlin, 1971; https://www.amazon.de/Einf%C3%BChrung-mathematische-Methoden-Kybernetik-K%C3%A4mmerer/dp/B00DE1BBP0
19. B. Schmidt - *Welt der Physik Meldung der Deutschen Physikalischen Gesellschaft* https://www.weltderphysik.de/gebiet/universum/news/2011/nobelpreis-fuer-physik-2011-geht-an-perlmutter-schmidt-und-riess/
20. A. Schopenhauer - *Die Welt als Wille und Vorstellung*; 1890 Brockhaus-Verlag Leipzig p.14; https://www.lernhelfer.de/sites/default/files/lexicon/pdf/BWS-DEU2-0958-03.pdf
21. L. D. Landau u. E. M. Lifschitz - *Lehrbuch der Theoretischen Physik Bd.II Klassische Feldtheorie;* p.4; Akademieverlag Berlin 1966; https://www.zvab.com/buch-suchen/titel/lehrbuch-der-theoretischen-physik/autor/landau-lifschitz
22. B. Riemann - *Über die Hypothesen, welche der Geometrie zu Grunde liegen ;* https://www.emis.de/classics/Riemann/Geom.pdf

23. H. Minkowski *Das Relativitätsprinzip*.; In: Annalen der Physik. 352, Nr. 15, 1907/1915, p. 927–938; https://de.wikisource.org/wiki/Das_Relativit%C3%A4tsprinzip_(Minkowski)

3. **The Observer**

1. I. Prigogine u. I. Stenglers - *Dialog mit der Natur, Neue Wege des wissenschaftlichen Denkens;* R. Pieper Verlag München,Zürich 1982 p.47; https://www.amazon.de/Dialog-Natur-Neue-naturwissenschaftlichen-Denkens/dp/3492030866
2. I. Kant - *Die Kritik der reinen Vernunft;* http://www.gutenberg.org/ebooks/6343
3. J. Glauberg - *Logica vetus et nova.* (1654), p. 320; . https://www.abebooks.com/servlet/BookDetailsPL?bi=17658885566&searchurl=tn%3Dlogica%2Bvetus%2Bnova%26sortby%3D17&cm_sp=snippet-_-srp1-_-title2
4. P. Marmet - *Absurdities in Modern Physics: A Solution;* https://www.newtonphysics.on.ca/heisenberg/
5. W..Heisenberg - Quantentheorie und Philosophie; p. 62 -75 Reclam Universalbibliothek Nr.9948, ISBN 978-3-15009948-3; https://www.amazon.de/Quantentheorie-Philosophie-Reclams-Universal-Bibliothek-Heisenberg/dp/315009948X
6. E. Schrödinger - *Schrödingers Katze* 1935; https://www.leifiphysik.de/atomphysik/quantenmech-atommodell/versuche/schroedingers-katze-ein-gedankenexperiment
7. WIKIPEDIA - Conway's Game of Life; https://en.wikipedia.org/wiki/Conway%27s_Game_of_Life
8. K. Kratky and F. Wallner - *Grundprinzipien der Selbstorganisation*, Wissenschaftliche Buchgesellschaft Darmstadt ; https://lux.leuphana.de/vufind/Record/025864815
9. L. Smolin – *Three Roads to Quantum Gravity;* 2001 Basic Books ISBN-13:978-0-465-07836-3; http://alpha.sinp.msu.ru/

~panov/LibBooks/SMOLIN/Lee_Smolin - Three_Roads_to_Quantum_Gravity-Basic_Books(2002).pdf

10. W. Heisenberg - *Physics and Philosophy, the Revolution in Modern Science,* New York, Harper and Row, 1966, 213 p. https://www.amazon.com/Physics-Philosophy-Revolution-Modern-Science/dp/0061209198
11. P. Marmet - *Absurdities in Modern Physics: A Solution;* https://www.newtonphysics.on.ca/heisenberg/
12. A. Uhlman - *Das Plancksche Wirkungsquantum hundert Jahre danach;* http://www.physik.uni-leipzig.de/~uhlmann/PDF/Uh00f_p.pd
13. P.-M. Robitaille and St. Crothers - “The Theory of Heat Radiation” Revisited: A Commentary on the Validity of Kirchhoff's Law of Thermal Emission and Max Planck’s Claim of Universality; https://www.researchgate.net/publication/280042096_The_Theory_of_Heat_Radiation_Revisited_A_Commentary_on_the_Validity_of_Kirchhoff's_Law_of_Thermal_Emission_and_Max_Planck's_Claim_of_Universality
14. St. Crothers - *A Nobel Laureate Talking Nonsense: Brian Schmidt, a Case Study;* http://vixra.org/pdf/1507.0130v1.pdf
15. WIKIPEDIA - Talk about *Plasmakosmologie;* https://en.wikipedia.org/wiki/Talk:Plasma_cosmology
16. G. Wunsch – *Karl Küpfmüller Wegbereiter der modernen Systemtheorie;* https://doi.org/10.1515/FREQ.1997.51.9-10.218
17. G. Wunsch – *Geschichte der Systemtheorie;* WTB-Reihe - Akademieverlag Berlin 1985; https://www.amazon.de/Geschichte-Systemtheorie-Dynamische-Systeme-Prozesse/dp/3486295314
18. A. L. Peratt - Advances in Numerical Modeling of Astrophysical and Space Plasma, APSS 242, 1997; http://plasmauniverse.info/downloads/AdvancesI.pdf
19. M. Livio - *Ist Gott ein Mathematiker?: Warum das Buch der Natur in der Sprache der Mathematik geschrieben ist;* https://www.amazon.de/Ist-Gott-ein-Mathematiker-geschrieben/dp/3423348003

20. W. Schmidt - *Die natürliche Selektion der Theoretischen Physiker* ; DPG-Didaktik-Tagungsband 1988, p. 593 - 599. Hrsg.: Prof. Dr. W. KUHN, Gießen; http://www.ekkehard-friebe.de/SCHMIDT.HTM

4. **On the conceptual world of physics**

1. R. Laughlin - *Der Urknall ist nur Marketing;* Interview im Spiegel 1/2008 http://magazin.spiegel.de/EpubDelivery/spiegel/pdf/55231886
2. P. p. Laplace: *Essai philosophique sur les probabilités.* Courcier, Paris 1814, English https://books.google.de/books?id=5vPxBwAAQBAJ&pg=PP8&lpg=PP8&dq=2.+P.+p.+Laplace:%C2%A0Essai+philosophique+sur+les+probabilit%C3%A9s.,eng+deutsch&source=bl&ots=2l2gNKAr94&sig=1WobzG5jatQo4oYcjIyn-bqGbil&hl=en&sa=X&ved=2ahUKEwiXqdqS9fDbAhUDC-wKHSf7BMoQ6AEwB3oECAEQUQ#v=onepage&q=2.%20P.%20p.%20Laplace%3A%C2%A0Essai%20philosophique%20sur%20les%20probabilit%C3%A9s.%2Ceng%20deutsch&f=false deutsch O. Höfling: Physik. Band II Teil 1, Mechanik, Wärme. 15. Edition. Ferd. Dümmlers Verlag, Bonn 1994, ISBN 3-427-41145-1 ; https://www.booklooker.de/B%C3%BCcher/Angebote/autor=H%C3%B6fling&titel=Physik+Band+II+Teil+1
3. B. Eckard – *Chaos;* p. Fischer Verlag 2004; ISBN-10:359615569X; https://www.amazon.de/Fischer-Kompakt-Chaos-Bruno-Eckhardt/dp/359615569X
4. I. Velikovsky - *Welten im Zusammenstoß;* Julia White Publishing; www.julia-white.com
5. I. Velikovsky - *Sternengucker und Totengräber* ;Julia White Publishing; www.julia-white.com

6. *Korrespondenz* Velikovsky -*Einstein;* http://www.supernatural-research.com/2010/07/albert-einsteins-final-letter-to-velikovsky/
7. L. Smolin - *Three Roads to Quantumgravity;* https://www.amazon.de/Three-Roads-Quantum-Gravity-Smolin/dp/0465094546
8. L. Smolin - *The Trouble with Physics;* https://www.amazon.com/Trouble-Physics-String-Theory-Science/dp/061891868X
9. G. Greiter - *Die 4 Grundkräfte der Physik*; http://greiterweb.de/spw/Grundkraft-der-Physik.htm
10. M. Hüfner - *Forschungslogik und Gravitation;* http://mugglebibliothek.de/katalog.htm
11. St. Hawking - *A Brief History of Time; Chapter5: Elementary Particles and the Forces of Nature;* https://www.amazon.com/Brief-History-Time-Stephen-Hawking/dp/0553380168
12. A. Einstein – *Zur Elektrodynamik bewegter Körper;* http://users.physik.fu-berlin.de/~kleinert/files/1905_17_891-921.pdf
13. P. Marmet – Fundamental Nature of Relativistic Mass and Magnetic Fields. ; http://www.newtonphysics.on.ca/magnetic
14. J. de Climont - *Eine Folge des Rowland-Effekts - Das intrinsische Feld des Elektrons ist kein Dipol;* http://editionsassailly.com/drehende_Leitern.pdf
15. P. Marmet - Natural Length Contraction Mechanism Due to Kinetic Energy; https://www.newtonphysics.on.ca/kinetic/
16. R. Kohlrausch und W. E. Weber, - *Ueber die Elektricitätsmenge, welche bei galvanischen Strömen durch den Querschnitt der Kette fließt* ; Annalen der Physik, 99, pg 10 (1856); https://onlinelibrary.wiley.com/doi/abs/10.1002/andp.18561750903
17. R. Haumann - *Die Physik des Nichts* Eigenverlag Viaveto.de
18. P. Marmet - *A New Non-Doppler Redshift* ; http://www.newtonphysics.on.ca/hubble
19. J. Eisenstaedt u. M. Combes- Arago and the Speed of Light (1806-1810): An Original Manuscript, a New Analysis; https://www.cairn-int.info/article-E_RHS_641_0059--arago-and-the-speed-of-light-1806-1810-a.htm

20. H. Fizeau, *(1860). - On the Effect of the Motion of a Body upon the Velocity with which it is traversed by Light. Philosophical Magazine. 19: 245–260.* https://en.wikisource.org/wiki/On_the_Effect_of_the_Motion_of_a_Body_upon_the_Velocity_with_which_it_is_traversed_by_Light
21. D. Miller - The Ether-Drift Experiment and the Determination of the Absolute Motion of the Earth; Reviews of Modern Physics, Vol.5(2), p.203-242, July 1933; http://adsabs.harvard.edu/abs/1933RvMP....5..203M
22. P. Marmet - The Overlooked Phenomena in the Michelson-Morley Experiment; http://www.newtonphysics.on.ca/michelson/
23. G. W. Hammar - *The Velocity of Light Within a Massive Enclosure;* Phys. Rev. 48, 462–463 (1935); https://journals.aps.org/pr/abstract/10.1103/PhysRev.48.462.2
24. B. P. Abbott u.a.- Observation of Gravitational Waves from a Binary Black Hole Merger; Phys. Rev. Lett. 116, 061102 ; https://journals.aps.org/prl/abstract/10.1103/PhysRevLett.116.061102
25. W. Orlov – Zirp von LIGO; http://walter-orlov.wg.am/zilch_von_ligo/
26. H. Barkhausen – Einführung in die Schwingungslehre; https://www.amazon.de/Schwingungslehre-Anwendungen-mechanische-elektrische-Schwingungen/dp/B00F8PY8UK
27. C. E. Shannon - *A Mathematical Theory of Communication.* In: *Bell System Technical Journal.*Vol 27, No. 3, 1948, p. 379–423, http://math.harvard.edu/~ctm/home/text/others/shannon/entropy/entropy.pdf
28. L. Susskind - *The Black Hole War*; p.133 ISBN 978-0-316-01640-7(hc)/978-0-316-01641-4(pb); https://www.amazon.com/Black-Hole-War-Stephen-Mechanics/dp/0316016411
29. B. Gaede – *Why God Does'nt Exist;* http://www.youstupidrelativist.com/

30. B. Mandelbrot - *Die fraktale Geometrie der Natur;* https://www.amazon.de/Fraktale-Geometrie-Natur-Beno%C3%AEt-Mandelbrot/dp/3764326468#reader_3764326468

5. The Microcosm

1. B. Greene – *Der Stoff, aus dem der Kosmos ist;* https://www.amazon.de/Stoff-aus-dem-Kosmos-ist/dp/3442154871
2. H. Niedderer & J. Petri - *Mit der Schrödinger-Gleichung vom H-Atom zum Festkörper - 4.4.6 Die experimentelle Bestimmung von Atomradien;* http://www.idn.uni-bremen.de/pubs/Niedderer/2000-QAP-JP-b.pdf
3. *Kernphysik -Grundlagen ;* https://www.leifiphysik.de/kern-teilchenphysik/kernphysik-grundlagen/versuche/ermittlung-der-kernradien-durch-streuung
4. R. Pohl u.a . - *Protonen-Radius neu vermessen* ; 2010 *Nature* ; 466, 213-216; https://internetchemie.info/news/2010/jul10/protonen-radius-neu-vermessen.php
5. F. Heiße u.a. - *Das Proton ist leichter als gedacht*, Phys. Rev. Lett. 119, 033001 – Published 18 July 2017 Science.de http://www.scinexx.de/wissen-aktuell-21680-2017-07-21.html
6. A. Unzicker - *The Higgs Fake: How Particle Physicists Fooled the Nobel Committee;* Kindle Edition 2013; https://www.amazon.com/Higgs-Fake-Particle-Physicists-Committee/dp/1492176249
7. J. Bleck-Neuhaus - *Elementare Teilchen --Von den Atomen über das Standardmodell bis zum Higgs-Boson* Springerverlag 2013; https://www.springer.com/de/book/9783642325786
8. A. Schopenhauer - "Die Welt als Wille und Vorstellung", 4 Bücher nebst einem Anhange, der die Kritik der Kantischen Philosophie enthält (Google eBook) S.51] https://de.scribd.com/doc/256236653/Schopenhauer-Kritik-Der-Kantischen-Philosophie-Backup
9. G. Klaus u. M. Buhr – *Philosophisches Wörterbuch Positivismus* VEB Bibliographisches Institut Leipzig 1966 ; https://

www.amazon.de/Philosophisches-W%C3%B6rterbuch-Klaus-Buhr/s?ie=UTF8&page=1&rh=i%3Aaps%2Ck%3APhilosophisches%20W%C3%B6rterbuch%20%28Klaus-Buhr%29

10. Sha Yin Yue - *On the Radius of the Neutron, Proton, Electron and the Atomic Nucleus* ; http://www.gsjournal.net/old/physics/yue.pdf.
11. *H. J. Giels – Experimentelle Untersuchungen der Radiusdifferenzen zwischen Protonen- und Neutronendichteverteilungen ...; Kernforschungzentrum Dortmund 1975; https://publikationen.bibliothek.kit.edu/200008880/3811426*
12. W. Gerlach, O. Stern - *Der experimentelle Nachweis des magnetischen Moments des Silberatoms,* Zeitschrift für Physik, 8, 110-111 (1921) chargeable: https://link.springer.com/article/10.1007%2FBF01329580
13. A. Einstein, W. J. de Haas, *Experimenteller Nachweis der Ampereschen Molekularströme*, Deutsche Physikalische Gesellschaft, Verhandlungen **17**, pp. 152–170 (1915); https://archive.org/stream/verhandlungen00goog#page/n167/mode/2up
14. J. de Climont: - *Eine Folge des Rowland-Effekts: Das intrinsische Magnetfeld des Elektrons ist kein Dipol;* http://docplayer.org/27453862-Eine-folge-des-rowland-effekts-das-intrinsische-magnetfeld-des-elektrons-ist-kein-dipol.html
15. S. Ulmer, et others - *Observation of Spin Flips with a Single Trapped Proton*; Phys. Rev. Lett. 106, 253001 – Published 20 June 2011; https://journals.aps.org/prl/abstract/10.1103/PhysRevLett.106.253001
16. A. Sommerfeld - *Optik Reihe Vorlesungen über Theoretische Physik;* Vol IV Akademische Verlagsgesellschaft Geest Porzig Leipzig 1964. https://www.amazon.de/Vorlesungen-%C3%BCber-Theoretische-Physik-Optik/dp/3871443778

17. St. Hawking - *A Brief History of Time; pp.*70-71; https://www.amazon.com/Brief-History-Time-Stephen-Hawking/dp/0553380168
18. W..Heisenberg - *Philosophie der Quantenmechanik;* Reclam Universalbibliothek Nr.9948, ISBN 978-3-15009948-3; https://www.amazon.de/Quantentheorie-Philosophie-Reclams-Universal-Bibliothek-Heisenberg/dp/315009948X
19. WIKIPEDIA – *Atomkern*;https://de.wikipedia.org/wiki/Atomkern#Kernmodelle
20. R. A. Alpher, H. Bethe u. G. Gamow - *The Origin of Chemical Elements;* Physical Review Vol73 No 7 1948 https://journals.aps.org/pr/pdf/10.1103/PhysRev.73.803
21. G. Lemaître - *The Primeval Atom: An Essay on Cosmogony;* https://science.sciencemag.org/content/116/3021/574.1
22. N. Bohr - *Neutron capture and nuclear constitution;* Nature Vol. 137, 344 (1936); https://www.nature.com/articles/137344a0
23. WIKIPEDIA - *Tröpfchenmodell;* https://de.wikipedia.org/wiki/Tr%C3%B6pfchenmodell
24. M. Hüfner - *Von Magiern,E=mc² und vom Kosmos Anlage 2 Zur Entwicklung von Galaxien aus ihrem Wasserstoffbrennen;* http://mugglebibliothek.de/huefner.htm
25. C. Johnson - *Nuclear Physics May be Fairly Simple;* http://mbsoft.com/public4/nuclei7.html
26. A. Unsöld - *Über die Temperatur der Sonnenkorona* ; Zeitschrift für Astrophysik, 1960 Vol. 50, p.48; http://adsabs.harvard.edu/full/1960ZA.....50...48U
27. W. Gerlach, O. Stern; *Der experimentelle Nachweis der Richtungsquantelung im Magnetfeld*; Zeitschrift für Physik December 1922, Volume 9, Issue 1, pp 349–352 https://link.springer.com/article/10.1007%2FBF01326983
28. A. Einstein, & W. J. Haas, - *Experimental proof of the existence of Ampère's molecular currents,* in: KNAW, Proceedings, 18 I,

1915, Amsterdam, 1915, pp. 696-711; http://www.dwc.knaw.nl/DL/publications/PU00012546.pdf

29. E. W. Schpolski - *Atomphysik Bd.II §288 Das Neutrino;* Deutscher Verlag der Wissenschaften Berlin 1962; https://www.antikvarium.hu/konyv/e-w-schpolski-atomphysik-i-ii-699821

30. C. Johnson - *Neutrinos Do Not Exist;* http://mb-soft.com/public4/neutrino.html

31. H. J. Wollersheim - *Direct and compound nucleus reaction*; Simon Fraser University 2011; https://web-docs.gsi.de/~wolle/TELEKOLLEG/KERN/LECTURE/Fraser/L24.pdf

32. T. A. Brun - *The Stern-Gerlach Experiment and Spin* ;http://www-bcf.usc.edu/~tbrun/Course/lecture02.pdf

33. T. E. Phipps, J. B. Taylor: - *The Magnetic Moment of the Hydrogen Atom.* Physical Review Band 29 (1927) p. 309–320

34. WIKIPEDIA - *Das Elektron;* https://de.wikipedia.org/wiki/Elektron

35. J. de Climont - *Eine Folge des Rowland-Effekts: Das intrinsische Magnetfeld des Elektrons ist kein Dipol.;* http://editionsassailly.com/drehende_Leitern.pdf

36. E. Kaal - The Proton-Electron Atom — A Proposal for a Structured Atomic Model ; https://www.thunderbolts.info/wp/2017/01/22/eu2017-speakers/ und http://www.everythingselectric.com/tag/edwin-kaal/

37. L. M. Brown - *The Idea of the Neutrino;* https://ddd.uab.cat/pub/ppascual/ppascualapu/ppascualapu_41_001@benasque.pdf

38. H. Ebert - Physikalisches Taschenbuch; Friedrich Vieweg Braunschweig 1967 New Edition; http://www.springer.com/de/book/9783528084172

39. J. p. Coursey, u.a. ; Atomic Weights and Isotopic Compositions for All Elements NIST Physical Measurement Laboratory ;

https://physics.nist.gov/cgi-bin/Compositions/stand_alone.pl?ele=&ascii=html

40. Thunderbolts Project TM - *EU2017 Future science;* Conferen in Phoenix 2017; https://www.thunderbolts.info/wp/2017/01/22/eu2017-homepage-2/

41. F. Bosch, i.o. - Observation of Bound-State β−Decay of Fully Ionized ^{187}Re: ^{187}Re−^{187}Os Cosmochronometry; Phys. Rev. Lett. 77, 5190 – 23 Dez. 1996; https://2014.f.a0z.ru/04/06-3430949-physrevlett.77.5190-1996.pdf

6. The Macrocosm

1. T. Davis u. Ch. Lineweaver - *Expanding Confusion:*common misconceptions of cosmological horizons and the superluminal expansion of the universe; 2003; https://arxiv.org/pdf/astro-ph/0310808.pdf#page=3
2. A. Einstein - *Die Grundlage der Allgemeinen Relativitätstheorie;* Annalen der Physik No7 forth Serie Vol.49 1916; https://onlinelibrary.wiley.com/doi/pdf/10.1002/andp.19163540702
3. M. Hüfner - *Einsteins dunkle Seite;* http://blog.mugglebibliothek.de/#post13 original: http://www.orgonelab.org/miller.htm
4. D. Miller - *The Ether-Drift Experiment and the Determination of the Absolute Motion of the Earth;* Reviews of Modern Physics, Vol.5(2), p.203-242, July 1933.
5. W. Thornhill – The long path to Understanding the Gravity; *https://www.youtube.com/watch?v=YkWiBxWieQU*
6. F. W. Dyson, u,a - *A determination of the deflection of light by the sun's gravitational field, from observations made at the total eclipse of May 29, 1919;* http://rsta.royalsocietypublishing.org/content/220/571-581/291
7. WIKIPEDIA – Jesuitenschule; https://de.wikipedia.org/wiki/Jesuiten#Der_Orden_als_Bildungsinstitution
8. W. Johnes - The ordinances of menu ; http://archive.org/stream/institutesofhind00manu#page/2/mode/2up

9. G. Lemaître - *Un univers homogène de Masse constante et de rayon croissant, rendant compte de la vitesse radiale des nébuleuses extra-galactiques.* Provided by the NASA Astrophysics; http://www-history.mcs.st-andrews.ac.uk/Biographies/Lemaitre.html und G. Lemaître:*Expansion of the universe, A homogeneous universe of constant mass and increasing radius accounting for the radial velocity of extra-galactic nebulae;* .Monthly Notices of the Royal Astronomical Society, Band 91, März 1931, p. 483–490, http://adsabs.harvard.edu/full/1931MNRAp..91..483L
10. G. Lemaître: - *Un Univers homogène de masse constante et de rayon croissant rendant compte de la vitesse radiale des nebuleuses extra-galactiques*. In: Annales de la Société Scientifique de Bruxelles, A47, 1927, p. 49–59.
11. G. Lemaître - *The Primeval Atom, An Essay on Cosmogony* ; D. Van Nostrand Co., New York, 1950, http://iloapp/mugglebibliothek.de/blog/blog?ShowFile6doc=152229866.pdf.
12. M. Bartusiak - *The Day We Found the Universe;* (2009) http://publicism.info/science/universe_1/16.html
13. A. Friedmann - Über die Krümmung des Raumes; Zeitschrift für Physik 10 (1): p. 377 – 386. (1922). https://link.springer.com/article/10.1007%2FBF01332580
14. J. J. O'Connor und E. F. Robertson:- Lemaître ; University of St. Andrews. Georges Henri-Joseph-Edouard. http://www-history.mcs.st-andrews.ac.uk/Biographies/Lemaitre.html
15. The Einstein Velikovsky Correspondence https://www.-varchive.org/cor/einstein/
16. S. Singh - *Big Bang;* HarperCollins UK.(2010). p. 362. ISBN 9780007375509.https://www.amazon.com/Big-Bang-Universe-Simon-Singh/dp/0007162219
17. E. T. Koch - *Von Christi Händen zu einem Urzustand Energiekonzentration: Eine Suche nach Annäherung von Schöp-*

fungsglaube und Naturwissenschaft am Beginn des 21.Jahrhunderts http://www.amazon.de/gp/search?index=books&linkCode=qs&keywords=9783839157756
18. Brief an Henry Cavendish, quoted from https://en.wikipedia.org/wiki/Dark_star_(Newtonian_mechanics)
19. P. Laplace - *The System of the World ;* Translation from French London 1809 https://archive.org/stream/system-world01laplgoog#page/n0/mode/2up
20. Max Planck Gesellschaft – *Black holes theorized in the 18th century;* 4,2017; https://phys.org/news/2017-04-black-holes-theorized-18th-century.html
21. St. Hawking, - *Particle creation by black holes,* Commun. Math. Phys. 43 (1975), 199–220; https://projecteuclid.org/download/pdf_1/euclid.cmp/1103899181
22. WIKIPEDIA *-Sloan Digital Sky Survey Project;* https://en.wikipedia.org/wiki/Sloan_Digital_Sky_Survey
23. Zooinverse - *Galaxy Zoo* https://www.zooniverse.org/projects/zookeeper/galaxy-zoo/
24. M. Odenwald - *Wissenschaftlicher Durchbruch - Schwarzes Loch M87: Was auf dem Sensationsfoto zu sehen ist;* https://www.focus.de/wissen/weltraum/odenwalds_universum/wis-senschaftliche-sensation-foto-von-schwarzem-loch-m87_id_10576481.html
25. WIKIPEDIA - *Schwarzschild - Metrik* https://de.wikipedia.org/wiki/Schwarzschild-Metrik
26. P. Marmet - *Einstein's Theory of Relativity versus Classical Mechanics* - https://www.newtonphysics.on.ca/einstein/
27. L. Susskind - *The Black Hole War, my battle with Stephen Hawking to make the World save for Quantum Mechanics,* Hachette Book Group; ISBN 978-0-316-01640-7(hc) / 978-0-316-01641-4(pb);https://www.amazon.com/Black-Hole-War-Stephen-Mechanics/dp/0316016411
28. L. Smolin - The *Trouble with Physics- The Rise of String Theory, The Fall of a Science, and What Comes Next;* Mariner

Book 2007, ISBN-13 978-0-618-55105-7 https://www.amazon.-com/Trouble-Physics-String-Theory-Science/dp/061891868X

29. R. Vaas - *Warum Stephen Hawking seine Wette verlor,* Wissenschaft.de ,https://www.wissenschaft.de/astronomie-physik/warum-stephen-hawking-seine-wette-verlor/
30. H. Ratcliffe – from *Die Welt nach Hawkings Willen, das verschwundene Loch in unseren Socken; Socks Chapter8* https://www.facebook.com/notes/hilton-ratcliffe/socks-chapter-8-the-world-according-to-hawking/10157087850989179/
31. Quote St. Hawking: https://www.linkedin.com/feed/update/urn%3Ali%3Aactivity%3A6406745903012593665/?midToken=AQEUYv8zBvFiAg&trk=eml-email_notification_single_shared_by_your_network_01-notifications-1-hero%7Ecard%7Efeed&trkEmail=eml-email_notification_single_shared_by_your_network_01-notifications-1-hero%7Ecard%7Efeed-null-7y1n7b%7Ejhpuomd4%7Ehb-null-voyagerOffline&lipi=urn%3Ali%3Apage
32. St. Hawking - "Information Preservation and Weather Forecasting for Black Holes"; https://arxiv.org/abs/1401.5761
33. N. Lossau - *Nachweis einer zweiten Gravitationswelle geglückt;* https://www.welt.de/wissenschaft/article152292984/Nachweis-einer-zweiten-Gravitationswelle-gegluecktu.html

7. **Outlook in a Intergalactic World**

1. The StarChild Team *Does the Sun move around the Milky Way? Febr.2000 https://starchild.gsfc.nasa.gov/docs/StarChild/questions/question18.html*
2. J. DeMeo - *Dayton Miller's Ether-Drift Experiments: A Fresh Look*; http://www.orgonelab.org/miller.htm

3. I. Velikovsky – *Before the Day Breaks - The Einstein Velikovsky Correspondence* http://www.varchive.org/bdb/main.htm
4. I. Velikovsky – *Welten im Zusammenstoß;* https://www.amazon.de/Welten-im-Zusammenstoss-Immanuel-Velikovsky/dp/3934402917
5. A. Egeland u.W. J. Burke (2005). *Kristian Birkeland – The First Space Scientist. Springer. ISBN 1-4020-3293-5.;* https://www.springer.com/de/book/9781402032936
6. H. Alfvén - *Cosmic Plasma;* Astrophysics and Space Science Library, Vol. 82 (1981) Springer Verlag. ISBN 90-277-1151-8 ; https://www.springer.com/west/home/physics?SGWID=4-10100-22-33598597-0
7. H. Alfvén - *Cosmology, Myth or Science?* J. Astrophys. (1984)5 p.78-95 http://adsabs.harvard.edu/abs/1984JApA....5...79A
8. H. Alfvén - *Cosmogony as an Extrapolation of Magnetospheric Research;* (1984) http://articles.adsabs.harvard.edu/cgi-bin/nph-iarticle_query?1984SSRv...39...65A&data_type=PDF_HIGH&type=PRINTER&filetype=.pdf
9. A. Peratt -. *Evolution of the Plasma Universe: II. The Formation of Systems of Galaxies* . IEEE Trans. on Plasma Science. (1986) PS-14: 763–778. ISSN 0093-3813.; http://public.lanl.gov/alp/plasma/downloadsCosmo/Peratt86TPS-II.pdf
10. E. Lerner - *The Big Bang never Happened* ; Vitage Books New York (1991), ISBN 0-679-740449-X (pb); https://www.amazon.de/Big-Bang-Never-Happened/dp/0812918533
11. D. Talbott – *Die Entdeckung der elektrischen Sonne Teil 1* ; http://mugglebibliothek.de/EU/sonne1.htm
12. A. Kriegl - *Differentialgeometrie;* https://www.mat.univie.ac.at/~kriegl/Skripten/diffgeom.pdf
13. M. Torrilhon - *Zur Numerik der idealen Magnetohydrodynamik;* http://www.mathcces.rwth-aachen.de/torrilhon/Torrilhon_Dissertation2003.pdf

14. S. Lundquist - *On the Stabilty of Magneto-hydrostatic Fields,* Phys.Rev. 1951; v.83(2),307-311 ; https://journals.aps.org/pr/abstract/10.1103/PhysRev.83.307
15. D. Scott - *Consequences Of The Lundquist Model Of A Force-Free Field Aligned Current.* Prog. Phys., 2015, v. 10(1), 167–178.; http://www.ptep-online.com/2015/PP-41-13.PDF
16. H. Arp – *Seeing Red Redshift, Cosmology and Academic Science;* Apeiron 1998 ISBN 0-9683689-0-5; http://open sciences.org/books/physics-and-cosmology/seeing-red-redshifts-cosmology-and-academic-science
17. F. Duru u.a. - *A plasma flow velocity boundary at Mars from the disappearance of electron plasma oscillations* Icarus 26(2010) p.74-82; http://www-pw.physics.uiowa.edu/~dag/publications/2010_APlasmaFlowVelocityBoundaryAtMarsFromTheDisappearanceOfElectronPlasmaOscillations_ICARUp.pdf
18. B. Fleck, u.a.- *Four Years of SOHO Discoveries;* NASA http://sohowww.nascom.nasa.gov/publications/ESA_Bull102.pdf
19. D. A. Riechers u.a. - *Imaging the Molecular Gas in a z=3.9 Quasar Host Galaxy at 0.3" Resolution: A Central, Sub-Kiloparsec Scale Star Formation Reservoir in APM 08279+5255*; https://arxiv.org/abs/0809.0754
20. C. Cofield - *Farthest Galaxy Yet Smashes Cosmic Distance Record;* 2016; https://www.space.com/32150-farthest-galaxy-smashes-cosmic-distance-record.html
21. P. Marmet - *A New Non-Doppler Redshift* ; Physics Essays, Vol. 1, No: 1, p. 24-32, 1988 ; http://www.newtonphysics.on.ca/hubble/index.html
22. M. Hüfner - *Zur Entwicklung von Galaxien aus ihrem Wasserstoffbrennen* in *Von Magiern, E=mc² und dem Kosmos;*http://mugglebibliothek.de/huefner.htm

23. E. Bañados u.a. - An 800-million-solar-mass black hole in a significantly neutral Universe at a redshift of 7.5; https://www.nature.com/articles/nature25180
24. P. Marmet - *The Cosmological Constant and the Redshift of Quasars;* http://www.newtonphysics.on.ca/quasars/index.html
25. P. Silva u.a. -*A plasma focus driven by a capacitor bank of tens of joules;* REVIEW OF SCIENTIFIC INSTRUMENTS VOLUME 73, NUMBER 7; https://www.researchgate.net/publication/233916221_A_plasma_focus_driven_by_a_capacitor_bank_of_tens_of_joules
26. H. Alfvén - *On the Importance of Electric Fields in the Magnetosphere and Interplanetary Space;* Space Science Reviews, Volume 7, Issue 2-3, pp. 140-148 http://articles.adsabs.harvard.edu/cgi-bin/nph-iarticle_query?1967SSRv....7..140A&data_type=PDF_HIGH&whole_paper=YES&type=PRINTER&filetype=.pdf
27. E. Lerner - *The Big Bang Never Happened ;* p.249 https://www.amazon.de/Big-Bang-Never-Happened/dp/0812918533
28. H. Arp - *Atlas of Peculiar Galaxies*; https://ned.ipac.caltech.edu/level5/Arp/paper.pdf
29. SDSS-Katalog - http://www.sdss.org/dr12/
30. H. A. Bethe - *Energy Production in Stars;* Phys. Rev. 55, 434 – Published 1 March 1939; https://journals.aps.org/pr/abstract/10.1103/PhysRev.55.43
31. R. Kaiser - *Gas in Milchstrasse lässt Quasar flackern;* https://www.weltderphysik.de/gebiet/universum/news/2016/gas-in-milchstrasse-laesst-quasar-flackern/
32. C. Lintott u.a.- *Galaxy-Zoo-1*: Data Release of Morphological Classifications for nearly 900,000 galaxies; https://arxiv.org/abs/1007.3265
33. Sloan Digital Sky Survey https://dr14.sdss.org/optical/spectrum/view
34. H. H. Voigt: Abriss der Astronomie ; BI-Wiss.-Verlag; 5. Edition; 1991; https://www.amazon.de/Abriss-Astronomie-Hans-H-Voigt/dp/3860256408

35. SDSS Data Base - Release 7 - http://cas.sdss.org/dr7/de/sdss/
36. P. Marmet - *Discovery of H_2, in Space Explains Dark Matter and Redshift;* Science & Technology , Frühjahr 2000, p.5 - 7; http://www.newtonphysics.on.ca/hydrogen
37. A. Peratt - Simulating spiral galaxies; Sky and Telescope (ISSN 0037-6604), vol. 68, Aug. 1984, p. 118-122; http://adsabs.harvard.edu/cgi-bin/nph-bib_query?bibcode=1984S%26T....68..118P&db_key=AST&data_type=HTML&format=&high=45fffb02f208845
38. M.Camenzind - Spektren der Sterne - Harvard Spektralklassifikation ;2017; https://www.lsw.uni-heidelberg.de/users/mcamenzi/HD_Spektren.pdf
39. D. Scott - Birkeland Currents: A Force-Free Field-Aligned Model; Progress in Physics, Vol. 11 (2015) S. 167- 179 http://www.ptep-online.com/2015/PP-41-13.PDF
40. SDSS Database https://skyserver.sdss.org/dr14/en/help/docs/intro.aspx
41. A. Peratt - *The role of particle beams and electrical currents of the plasma universe;-* 1987, http://adsabs.harvard.edu/abs/1988LPB.....6..471P
42. H. Täger - *History of Electric Comet Theory: An Introduction;* https://www.thunderbolts.info/wp/2017/06/19/history-of-electric-comet-theory-an-introduction/
43. E. L. Lerner - *The Big Bang Never Happened ;*p. 188; Vintage Books New York 1991 ISBN 0-679-74049-X; https://www.amazon.de/Big-Bang-Never-Happened/dp/0812918533 und http://www.bigbangneverhappened.org/index.htm
44. H. v. Helmholtz: *Ueber einige Gesetze der Vertheilung elektrischer Ströme in körperlichen Leitern, mit Anwendung auf die thierisch elektrischen Versuche*. In: J. C. Poggendorff (Hrsg.): *Annalen der Physik und Chemie*. Dritte Reihe. Band 89. Verlag von Johann Ambrosius Barth, Leipzig 1853, p. 211–233,

45. E. Marsch, B. Inhester – *Helioseismologie und das Innere der Sonne* MPS Forschungsinformation 6/ 2005; https://www.mp-s.mpg.de/442607/12Helioseismologie-und-das-Innere-der-Sonne.pdf
46. *Das Periodensystem der Elemente online* hhttp://www.periodensystem-online.de/index.php? el=20&id=modify
47. R. E. Juergens *THE PHOTOSPHERE: IS IT THE TOP OR THE BOTTOM OF THE PHENOMENON WE CALL THE SUN?* Kronos Vol IV No4 1979; https://www.kronos- press.com/juergens/1979-photosphere-juergens.pdf
48. I. Langmuir,- *Proceedings, International Congress on…; General Electric Review No. 27 (1924) Physics, Como* (1927); *Journal of the Franklin Institute*, vol. 24, no. 3 (1932). See *Langmuir's Collected Works*, vol. 4, pp. 75, 140, and 179.
49. WIKIPEDIA- *Dissipative system;* https://en.wikipedia.org/wiki/Dissipative_system
50. NASA, Note of dec.13.1999: *The Day the Solar Wind Disappeared;* https://science.nasa.gov/science-news/science-at-nasa/1999/ast13dec99_1
51. H. Alfvén, *Electric Current Model of Magnetosphere* 1979; http://www.iaea.org/inis/collection/NCLCollectionStore/_Public/10/489/10489393.pdf
52. Cf. J. Kirk und W. Livingston, *Solar Physics* 3, 510 (1968). quoted by http://adsabs.harvard.edu/full/1993A%26A...279..599N
53. D. Scott Electric Solar Wind | EU2016, https://www.youtube.-com/watch?v=sFGb7NlUvgg
54. D. Scott The Electric Sky, https://www.amazon.com/Electric-Sky-Donald-Scott/dp/0977285111
55. M. Childs u. M. Clarage – *The SAFIRE-Project*; https://www.safireproject.com
56. H. Borchert - Die Globale Wärmeperiode wurde durch Sonnenaktivität verursacht und neigt sich dem Ende zu; http://www.umad.de/infos/downloads/die%20w%C3%A4rmeperiode%20neigt%20sich%20dem%20ende%20zu%202009.pdf

57. *CERN's CLOUD experiment provides unprecedented insight into cloud formation.;* CERN 2011 Press Release n°15
58. F. Bertola, u.a. *Counter rotating Stellar Disks in Early-Type Spirals:* NGC 3593 http://adsabs.harvard.edu/full/1996ApJ...458L..67B
59. H. Haken by Kratky u. Wallner - *Grundprinzipien der Selbstorganisation ;* https://www.amazon.de/Grundprinzipien-Selbstorganisation-Karl-W-Kratky/dp/3534109716
60. G. Pollack - *The Fourth Phase of Water: Beyond Solid, Liquid, and Vapor;* https://www.amazon.com/Fourth-Phase-Water-Beyond-Liquid/dp/0962689548
61. G. Pollack - *Weather and EZ Water – An Intimate Role of Separated Charge;* Vortrag auf der EU2017 in Phoenix https://www.youtube.com/watch?time_continue=597&v=KnwAUVN-hU0s
62. E. W. Schpolski *Atomphysik Bd.II ;* VEB Deutscher Verlag der Wissenschaften Berlin 1962 https://www.antikvarium.hu/konyv/e-w-schpolski-atomphysik-i-ii-699821
63. P. Günther - Einführung in die Plasmaphysik ; https://www.ipp.mpg.de/1166987/script_ws.pdf
64. G. Holman – Solar Flare Theory - What is a Solar Flare? https://hesperia.gsfc.nasa.gov/sftheory/flare.htm

8. **Other cosmic Puzzles**

1. Gold - *Rotating Neutron Stars and the Nature of Pulsars;* Nature **volume 221**, pages 25–27 (04 January 1969) https://www.nature.com/articles/221025a0
2. W. Hermsen, u.a. - *Simultaneous X-ray and radio observations of the radio-mode switching pulsar PSR B1822-09;* 13.Dez 2016, Cornell University Library; https://arxiv.org/abs/1612.04392

3. K. R. Healy u. A. L. Peratt - Radiation Properties of Pulsar Magnetospheres: Observation, Theory, and Experiment; Plasma Astrophysics and Cosmology pp 229-253 https://link.springer.-com/article/10.1007/BF0067802
4. F. C. Michel - Theory of pulsar magnetospheres; Rev. Mod. Phys. 54, 1 – Published 1 January 1982; https://journals.ap-s.org/rmp/abstract/10.1103/RevModPhys.54.1
5. D. Talbott - The Saturn Myth; https://www.amazon.de/Saturn-Myth-David-Talbott/dp/1979835489

Index